NUCLEAR ENERGY—
ITS PHYSICS
AND ITS SOCIAL CHALLENGE

NUCLEAR ENERGY—
ITS PHYSICS
AND ITS SOCIAL CHALLENGE

David Rittenhouse Inglis, *University of Massachusetts*

ADDISON-WESLEY PUBLISHING COMPANY
Reading, Massachusetts
Menlo Park, California · London · Don Mills, Ontario

This book is in the
ADDISON-WESLEY SERIES IN PHYSICS

Preface

The discovery of nuclear fission has considerably altered the prospects of mankind. Nuclear energy can prolong our power-based civilization or it can end it abruptly. The attendant radioactivity can cause disease or it can cure disease. Intelligent social choices must be based on technical as well as political understanding. For his survival man's intelligence is challenged as never before.

Among college students choosing courses of study and among citizens developing concern there is a cry for relevance to life. The technical understanding of the nature of nuclear reactors and bombs and appreciation of the political and diplomatic pressures associated with keeping them in their place are surely among the most relevant of topics. A modest acquaintance with the basic principles at work in the atoms and nuclei that activate the reactors and bombs is likewise relevant. Such relevance is the stuff of this book.

There has been talk in academic circles of the two cultures and of the gulf between them. One is the scientific culture, with all of its noble quest for new knowledge of the workings of the physical world and its pursuit of new applications of this knowledge for the presumed material benefit of mankind. The other is the humanistic culture, the various disciplines enlarging knowledge and understanding of man and his arts and of the interactions of man with man. Both are parts of what has come to be known as "the establishment," the established way that society has developed of doing things.

There is in a different sense and largely outside academic circles and even outside the establishment a third culture, an antiestablishment culture that is more popular among the present younger generation than it was with earlier ones. Highly critical of the most glaring shortcomings of the establishment to the point of talking of destroying it, scornful of its virtues in spite of being parasitically dependent on them, many of these young people have no interest in improving the way things are done within the establishment. A more serious shortcoming from their point of view is that they have little idea of what they would do to improve matters if their culture should succeed in taking charge of things. They profess keen interest in ecology and in preserving a natural environment. Yet they have little faith in knowledge passed on from the experience of others and would prefer to have each generation learn from its own experience, essentially forgoing the advantage humans have over the other animals in the ability to communicate. How they would handle various detailed problems if they were in charge—for example those of nuclear energy—remains a blank. Yet within or without the establishment, these problems will not just go away.

Between the three cultures, and particularly inside the two cultures within the establishment but in their fringes off toward the third, there is a large group of thoughtful people sincerely devoted to removing the shortcomings of the establishment and making it work better, in understanding the roots of the troubles and solving the more serious problems. Here lies the fertile ground for growing knowledge and constructive criticism that may keep our civilization viable and prevent its collapse in the wake of those that have gone before. This group can hopefully draw its strength from all three cultures—from youth of various degrees of concern and dissatisfaction as well as from scientists and humanists of longer experience and growing concern. To cope with such knotty problems as nuclear energy, all need to learn—the scientists from the humanities, the humanists from science, the youth from both and both from youth.

This volume attempts to bring to the reader, and particularly to the reader who is youthful at least in spirit, a piercing glimpse into both the scientific and humanistic aspects of nuclear-energy problems, including the problems of nuclear weapons, from the point of view of a scientist who professes enough concern that he may be given heed.

In academic circles, effort to acquire knowledge or understanding, and hopefully a bit of each, comes in a standard package size known as the semester course. This book is intended to be useful as the basis for such a package, and the material has been used in that way. The package is heaped to overflowing and the subject could be covered more thoroughly in two semesters.

The overriding aim of the book's organization is to provide the basis for understanding both the technical and the socio-political 'sides of the problems without prerequisite knowledge of either. This requires sufficient brevity that an instructor from either side of the fence may complain that his subject has been covered insufficiently to impart real understanding. If we succumb to the shibboleth that "a little knowledge is a dangerous thing," we will leave the problems to the self-interest of the experts. Or we will leave it at best to the few science students who will seek out the political problems or the humanist students who will study their physics thoroughly enough to know more than superficially the technology of which they may otherwise only glibly speak. Insofar as it is to be used as a textbook, the volume is aimed primarily at the general "arts" student who has little motivation toward understanding science, but needs the part directly relevant to nuclear-energy problems as a background for present and future political judgments. To such judgments the book also gives an introduction in some depth.

An understanding of the fission and fusion processes and the way they exploit both the light-nucleus and the heavy-nucleus end of the array of nuclei gives depth to a politically relevant judgment of the likelihood of similar miracles of science in the future. For this one must know the habits of protons and neutrons, which in turn requires knowing, for example, something about force and mass. The presentation of physical principles starts "from the beginning," helping the student acquire the general idea of needed concepts without the kind of lengthy drill considered essential in a standard physics course to develop skill in their use. Here and elsewhere the reader or an instructor may find supplementary sources useful, depending on the time that can be made available. Additional economy of effort is achieved by confining the concepts discussed to those leading rather directly to the treatment of reactors and bombs. An instructor with scientific background may be tempted to amplify the initial presentations to the extent of leaving little or no time in a short course for the social aspects later on. If so, he is urged to consider that in this context a little knowledge of both sides is apt to be more broadening for the student than a more thorough understanding of either.

The course on which this book is based has been taught three successive spring semesters at the University of Massachusetts (with no prerequisite other than high school algebra), first in one large section, then in a large general section plus a smaller section of those having already studied some science, and next as a general section and a smaller honors section.

Opinions may differ widely on how much basic physics is needed, or on how little may be meaningfully taught, as a useful background to understanding the workings of reactors and bombs sufficiently well to enrich political and economic discussions of their consequences. The first three chapters are something of a compromise in this respect, providing a brief statement and explanation of the most relevant principles needed without unduly postponing attention to the more timely topics that follow. Whether more or less preliminary background is needed may depend on the individual reader or on the astuteness of a class. One instructor may decide to skip this material and fill in what seems necessary in his own words or from another text. Another instructor may decide that it is not enough and may take time to be more thorough by using the appendices and other supplementary material as well. The format of text-plus-appendices is intended to leave flexibility in this respect. The usual discipline concerning units of force and mass (involving "slugs") is avoided without inconsistency by occasionally referring to mass in kilograms or to force in pounds. Of the material in the first chapters the least elementary is probably the explanation of the binding-energy curve in Chapter 3, which might be omitted with some groups though it is central to nuclear energy release. The independent reader may even decide to skip the first three chapters initially and refer back to them as need arises later on. These chapters, with somewhat less than half of the material in the associated appendices, represent about the level of presentation of physics that has been found suitable for the general section.

The reader, be he a student in a formal college course or a lifelong student of matters of interest and concern, may develop from these pages an interest in the future course of the "public atom." It is to be hoped that widespread well-informed citizen interest may through the democratic process guide national policy in a direction giving adequate weight to the long-term needs of society as contrasted with short-term gain.

The writer is grateful for having been aided in the development of understanding and views on these matters by a long succession of colleagues and associates, first at Los Alamos at the birth of these troublesome challenges, then at Johns Hopkins, over many years of discussions at the Argonne National Laboratory, in connection with *The Bulletin of the Atomic Scientists* and the Federation of American Scientists, in the Pugwash movement. Finally he has been happy to work among a group of colleagues at the University of Massachusetts. Of these Professor Allan Hoffman was the most directly involved with the course on the material of this book and has been helpfully critical of the manuscript.

Amherst, Massachusetts   D.R.I
September 1972

Contents

Chapter 9   Constraints on the Arms Race

INTRODUCTION Man's Growing Use of Power

In order to put in perspective man's very recent feat of releasing nuclear energy, let us look at the time scale on which energy has been variously harnessed and utilized.   About five billion years ago something tremendously violent happened in our corner of the universe to form the great variety of atomic nuclei that are the basis of our chemical elements. Locked within the nuclei are enormous amounts of energy. Almost that long ago the solar system with its planets, including the Earth, was formed. It was fairly recently, about five hundred million years ago, after the earth was teeming with life, that there was a prolific age known as the carboniferous era, lasting about a hundred million years. During this remote time, over much of the earth's surface trees grew rapidly and fell and decayed and were buried eventually to form coal. Thus the energy from the sun was converted into the energy of coal by photosynthesis in leaves. In the seas during those remote times small marine organisms multiplied, feeding on the microscopic flora flourishing in the sunlight and, dying, settled in huge beds that now are oil fields beneath the surface of the earth, partly disintegrated into pockets of natural gas.

Then, a very short time ago on this grand scale, about a million years ago, man evolved. For most of the time since then, his only power source was his muscle. In the prime of each man's life this amounts to about one-tenth horsepower. Perhaps ten thousand years ago he began to use fire for heat and domestic animals for power, about one horsepower per animal. Presumably not long after that he began to use the power of wind to propel his ships, and it was less than a thousand years ago that he made effective use of the power of wind and falling water to grind his grain.

Although before then a little coal was used for smelting, it was not until 200 years ago that appreciable amounts were mined and used. Until then, wood was essentially man's only fuel, wood that represents recent conversion of the sun's rays into fuel energy. The last 300 or 400 years have seen a remarkably rapid increase in man's understanding of the miracles of nature, and with this still-growing understanding has come a fruitful spirit of technological innovation. The steam engine between 200 and 100 years ago was powered mostly by wood, though it was part of the incentive for mining of coal in England. In the United States wood was plentiful and supplied about 90 percent of the fuel for power and heating as late as the year 1850, when the picturesque houses that still stand about us were all heated from the woodlot. After that the use of coal increased rapidly, and within the last 100 years the use of petroleum and natural gas was introduced and increased enormously until now, largely because of the automobile and gas heating, these constitute a major part of fuel consumption, particularly in the United States. (See Table 0–1.)

TABLE 0–1

Approximate relative fuel consumption
(as of about 1962)

	World	U.S.
	percent	
Coal	57	23
Oil	30	44
Natural gas	3	28
Hydroelectric	10	4
Wood	?	1

These percentages tell only part of the story. More important is the rapid increase in total use of power. In the United States the total annual use of power in 1960 was over ten times as much as it was in 1880. And 1880 is not so very prehistoric; the country was extensively industrialized, with many mills and factories and with powerful steam railroads. Our power consumption is expected to increase by another factor 2.5 or 3 by the year 2000, making it about 30 times as much as in

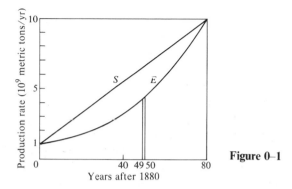

Figure 0-1

1880.* The total in 1960 was 1.7×10^9 tons of coal equivalent (of which 23 percent was actually coal).

Given the information that there was about a tenfold increase in 80 years, we might ask how this increase was distributed over the years. Did it follow a straight-line plot, like S in Fig. 0-1, or a gradually upward curving path like curve E? Line S represents a constant amount of growth year after year, just as much in the first year when the economy was small as in the last year when the economy was much larger. That would hardly be expected. Curve E is what is known as an exponential curve. It represents a growth by the same percentage year after year, in this case each year about 3 percent of the amount at the beginning of that year. (Those who know the language of calculus may recognize this as $(\text{Log}_e 10)/80 = 2.3/80 = 2.9$ percent, but one does not need to understand calculus to understand the nature of an exponential curve.) An exponential curve is simply a curve that gets steeper as it gets higher (or less steep as it gets lower), in proportion to its height. The steepness or slope represents the actual rate of increase, per year in this case.

The exponential curve seems like a more natural kind of increase, for it is reasonable that the amount of increase in use of fuel should depend to some extent on how much is already used. Without looking up historical statistics, it is a fair bet that the actual increase was much more nearly like curve E than line S. If the curve E is extended for another 40 years, from 1960 to 2000, at the same 3 percent per year increase, it goes up another factor 3.2, to 32 on the left-hand scale.

---

* Phillip Sporn, *Energy*, Pergamon Press, 1963.

The point of this discussion of the exponential curve is to emphasize in how short a time we have used most of the energy and most of the fuel that has ever been used by man, as well as to help guess future use. (Also, we are going to have other uses for the exponential curve before we are through with nuclear energy.) The area under curve E may be used to represent the total energy consumed, as may be seen by considering a vertical strip one year wide. Imagine drawing two vertical lines a year apart, for example, at the year marks 49 and 50, from the bottom of the graph up to the curve E. The height of the strip between them represents energy used per year. The width of the strip is one year. The area of the vertical strip is its height times its width, or energy per year times one year—that is, the energy used in that year. (The strip is cut off slantwise at the top, so it is not really rectangular, but that makes so little difference in figuring the area of a narrow strip that it doesn't matter.)

The total energy under curve E is the sum of all such strips, or the energy used in all 80 years. Now, the upturned shape of the graph is such that half of the area is in approximately the last 24 years (the shaded area in Fig. 0–2) as seems roughly reasonable by inspection. This is so even when the little "tail" that could be drawn off to the left of the graph is included, if we assume that the exponential curve applies also at earlier times. Thus in the short time of about a quarter century, we have used something like half of all the energy ever used by man.

Incidentally, the height of the curve also doubled, from 5 to 10, in those 24 years. That is another peculiarity of an exponential curve: the height of the curve doubles in the same time the area under the curve

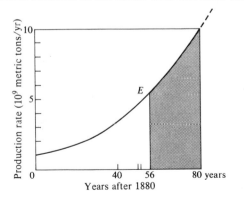

Figure 0–2

doubles. A rising exponential curve has a doubling time and for the overall rate of energy consumption in the United States this is roughly 24 years.

Because geological knowledge is incomplete, one cannot guess with certainty how long it will take for us, at this rapidly increasing rate, to use up the bounty of those many millions of years. Estimates in the past of imminent shortage have proved to be wrong because of unanticipated new discoveries of coal and oil and gas deposits. Particularly in the case of oil and gas, continually refined exploration, involving pushing out into the continental shelf under the edge of the oceans and into remote regions like Alaska, continues to make possible new discoveries. But this cannot go on forever. The United States tax structure is designed to encourage exploration and discourage import, as if intended to exhaust our national reserves as quickly as possible. The purpose of discouraging import is to help maintain a satisfactory balance of trade in spite of our other heavy imports (such as of metals of which we have already exhausted most of our internal natural resources) and in spite of expenditures abroad for military operations. These military operations thus cut into our domestic oil reserves in two ways, both by using a lot of oil and by aggravating the financial reason for discouraging oil import.

Coal is the most plentiful of the "fossil fuels," but gas and oil supplies are being consumed faster because they are more convenient to obtain. Though our growing economy still appears to be on an exponentially rising curve of fuel consumption, such a rise cannot go on forever. How it will level off and eventually decline is a matter of conjecture. A careful estimate of total coal deposits in the world, taking into account the fact that past assessments have usually underestimated the extent of new discovery, combined with a conjecture as to the probable rise and decline of consumption, lead to the curve of predicted world coal consumption shown in Fig. 0–3. It is not expected that consumption will follow a rapidly rising exponential curve until suddenly there is no more. Instead of a sudden and cataclysmic cutoff, there will probably be a gradual topping off and decline as deposits become harder to exploit. Note that consumption up to the present is represented by the small shaded area in the lower left corner, only about 2 percent of the total represented by the area under the complete curve. This puts us still on the exponentially rising part of the curve and allows

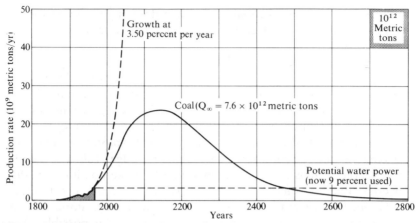

Fig. 0-3 Estimated course of world coal production, as compared with total hydroelectric potential. After M. K. Hubbert.[2]

for an increase of consumption by about a factor 10 before reaching a maximum in about the year 2100. After that there will be a gradual decline as deposits that are increasingly difficult to extract are exhausted.

The use of oil is expected to follow a similar bell-shaped curve on a

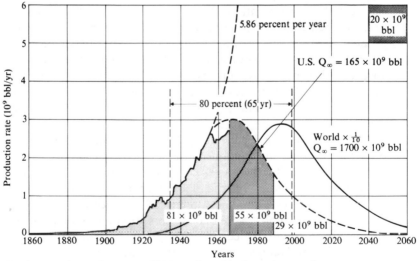

Fig. 0-4 Estimated course of U.S. and world oil production.[2]

much shorter time scale, as shown in Fig. 0–4 separately for the United States and for the world. Here the effect of United States tax policy in hastening depletion is evident in the earlier peaking of the United States curve about at the present time, 30 years ahead of the world curve. Exploring for new oil deposits involves expensive drilling of deep holes in the ground, guided by various technical soundings. In the United States the number of barrels of oil yield per foot of hole drilled declined from about 250 in 1935 to 35 in 1965, showing how rapidly it is getting harder to discover new deposits. Exploration of large areas has been most intense in the United States, and experience with discovery here along with geological study can be used to estimate what may be the ultimate production in other parts of the world Exhaustion of deposits here is expected earlier than elsewhere, but even on a worldwide basis oil and gas cannot last for many decades at the present rapidly increasing rate of consumption. Whether these estimates may prove to be pessimistic, as some previous estimates have been, remains to be seen.

The oil used to date amounts in heat equivalent to about half of the coal used to date. About 2 percent of the total coal deposits and about 3 percent of the total fossil fuel deposits have been used. But already there are serious questions about the environmental effects of the coal and other fossil fuels we are burning, not only the effects of loading the atmosphere with carbon compounds long locked in the earth but also the devastation of large areas of landscape by strip mining. Whether the environmental considerations will permit the exploitation of all the other 98 percent of the coal remains to be seen. Since this source of energy is potentially so important, research and development of efficient and more nearly pollution-free methods of burning fossil fuels should have the benefit of large-scale promotion such as would be possible with federal sponsorship. Such sponsorship has been bestowed on nuclear energy but not on fossil fuels. No single company, in an industry of a hundred or so companies, can afford the research needed.

Fossil fuels use solar energy from the remote past. Water power uses present-day solar energy that evaporates water from the seas to fall at higher altitude on land and flow down the streams and into the dammed reservoirs of electric generating plants. In some locations the amount of land that must be flooded and the size of dam that must be built to harness this falling water is reasonable in our present economy, in others not. Thus not all the potential of hydroelectric power is developed. The world total potential is about 2×10^9 kilowatts or 3.3×10^9 tons of coal equivalent per year. (Note that this is about twice the total annual United States energy consumption in 1960.) Of this total about 10 percent is in North America and 5 percent in western Europe. Of the world total, only about 9 percent is developed. Of the 9 percent actually developed, 57 percent is in western Europe and 23 percent in the United States.

In Fig. 0–3 the world total potential is indicated. It takes about 2000 years for this to equal the total amount of energy from coal available (the area under the straight line if extended out that far then being equal to the area under the coal-consumption curve). There is a serious problem of silting of reservoirs after one or a few centuries, after which they can no longer even out seasonal flow. The long-term potential will thus not be realized unless a solution to this problem can be found.

Tidal energy is much less promising. The total power of all the tides of the world is a bit larger than the hydroelectric potential, about 3×10^9 kilowatts, but only about 2 percent of this in special estuaries is considered to be potentially harnessable and 0.01 percent, or 300 megawatts, is actually harnessed in France. Geothermal power, the power from regions with hot springs, is another minor source. World-wide use of geothermal power, most of it in California and Italy, amounts to about 1000 megawatts, or about 0.05 percent of the world's potential water power. Such installations use up the stored heat and may last for only a century or so.

Nuclear energy, when and if it becomes an important contributor, will be important mainly as a producer of electric power. It is perhaps surprising how small a part of the total energy consumption we have been discussing is in the form of electric power. This is especially true in the United States where automotive transportation is so prevalent.

The first commercial electric power station (Edison's) went into service in New York City in 1882. It was of course small and inefficient, but grew rapidly, from supplying 400 80-watt lamps at first to 10^4 lamps and a maximum load of about 10^6 watts (1000 kilowatts or 1 megawatt or 7×10^5 horsepower) in about the first year. Rapid growth in the production and consumption of electric power has continued ever since, but while the total power consumption grew by a factor 10 between 1880 and 1960, the fraction of the fuel consumption converted to electricity grew from zero to only about 20 percent of the total (about 23 percent in 1970). The average efficiency of electric power plants is about 35 percent so only about 35 percent of this 20 percent, or about 7 percent of the total, emerges as electric power. In 1960, a convenient reference year, this amounted to 8×10^{11} kilowatt-hours per year (as one can verify from the total 1.7×10^9 tons of coal equivalent by use of the appropriate conversion factor from Appendix 13).

Modern generating plants are much more efficient than Edison's early one, requiring only about a tenth as much coal per kilowatt-hour of electric energy. Furthermore, lamps are more efficient, giving out much more light intensity per watt. Even with this efficiency, electricity remains a rather small part of our energy uses in spite of powering many appliances, machines, industries, and some transportation, particularly the elevators in high city structures.

All this combustion of fossil fuel, and even the evaporation from the sea that supplies hydroelectric plants, involves only the exterior or electronic part of the atom. What we have described is the power picture into which has recently entered the exciting prospect of running electric generating plants on the power locked in the inner nucleus of the atom. This is commonly called atomic power but is more properly nuclear power.

Power is an everyday idea. It denotes something about the ability to get things done fast or to provide heat and light. In our everyday dealings we need to know that it is sometimes measured in horsepower and sometimes in kilowatts. Though we use that word horsepower, nowadays we do not get much of our power from horses. We get most of it from burning, from the chemical behavior of molecules and their atoms. We are beginning to get a little of it from splitting the nuclei of heavy atoms, and the energy of the most powerful military threat hanging over the world depends on the combining or "fusion" of light nuclei to make heavier nuclei. These things are important to the economic basis of politics and to the social problems of war and peace. To understand them properly, so as to be able to make independent judgments without having to decide blindly which expert's word to take for each new technical problem as it comes up, we should develop some feeling for the basic nature of molecules and atoms and nuclei. Some may want to do this quite superficially; others may have the curiosity and patience to delve a little deeper.

Physics in all its depth and breadth is an abstruse subject, perhaps, but the basic principles and ideas important for energy production can be understood on a level that need not be very involved and can nevertheless be helpful as a background for weighing technical alternatives and for sound social and political judgments.

This discussion therefore turns now to a presentation of selected physical principles starting from "the very beginning" for the sake of those who have had no successful background in science at all. Some readers may prefer to skip some of the introductory material of the first three chapters and go on to the more specific discussion of reactors and bombs, planning to return to the introductory material as needed.

CHAPTER 1

Basic Physical Ideas

VELOCITY AND MOMENTUM

Positions are described relative to some reference point. To tell a friend how to find a place in the country, you start from some landmark he knows. A surveyor may describe a point as so many feet east and so many north from a bench mark or reference point. These distances may be called coordinates. When he locates the point on his map drawn to scale, the position may be represented by an arrow reaching from the reference point or "origin" to the point. The arrow is called a position vector and it has both a length and a direction—so many feet long, so many degrees east of north, for example.

The idea of a vector is useful because two vectors may be added together to form a third, but not in the same way that ordinary numbers are added. The length of the sum may be less than the sum of the lengths because the two vectors may have different directions. In Fig. 1-1, an object may be moved from the original position O to position A, and then from A to B, or the same result may be accomplished by moving it directly from O to B. Its first position at A is determined by the position vector reaching from O to A. This plus the change-of-position vector reaching from A to B is equal to the final position vector reaching from O to B. Any two vectors are added by forming such a triangle. The tail of the second vector is placed at the point of the first and the vector sum reaches from the tail of the first to the point of the second.

Velocity is a measure of how fast an object is changing its position, how fast it is getting from one place to another. In other words, velocity is the rate of change of position. It also has a magnitude and a direction and when drawn to scale may be represented by a vector, the velocity

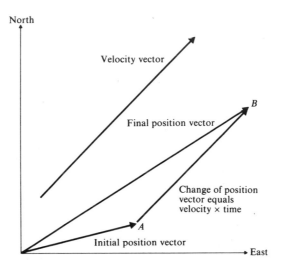

Fig. 1–1 When a body moves from *A* to *B*, its initial position plus its change of position is equal to its final position. Two vectors add in this way, with the tail of one on the point of the other, and their sum forms the third side of a triangle. If it moves at constant velocity (a vector), its change of position is its velocity times the time.

vector. We may for example represent 10 miles per hour as an inch on a piece of paper. Then the velocity of a car moving northeast at 30 miles per hour is represented on paper by an arrow 3 inches long pointing upward toward the right.

That car also has momentum. This is defined as its mass multiplied by its velocity, and it is also a vector. If the car has a mass of 1 ton, for example, its momentum is 30 ton-miles-per-hour toward the northeast. We can think of momentum as a sort of tendency to keep on moving in the same direction. If a minicar and a heavy truck both move at 30 miles per hour, the truck has a lot more momentum than the minicar, as can be appreciated if they hit something. A tree might be demolished while applying enough force to the truck to slow it down, but the force needed to stop the minicar could be applied without making much of a dent in the tree. Similarly, it takes a more powerful motor, developing a greater force, to accelerate the truck than the minicar up to 30 miles per hour in a given time and thus to create its greater momentum.

ACCELERATION AND FORCE

These are examples of the fact that force must be applied to change momentum, and force is defined as the rate of change of momentum. We have defined momentum as mass times velocity. The mass of the objects we are talking about remains constant, so a change of momentum is due to a change in velocity. Thus the force applied, or the rate of change of momentum, is the mass times the rate of change of velocity. The rate of change of velocity is called acceleration.

To repeat in equations what we have been saying:

Force = rate of change of momentum
   = rate of change of (mass times velocity)
   = mass times (rate of change of velocity)
   = mass times acceleration

or in briefer symbols,

$$F = ma.$$

Both force and acceleration have direction and are vectors.

A car moving along a straight road can be given a positive acceleration with the accelerator or a negative acceleration (deceleration) with the brakes, but as it rounds a curve without speeding up or slowing down it can be given a sideways acceleration with the steering wheel. In any of these cases the road applies a force to the car through friction on the tires, and the sideways force to supply the sideways acceleration is just as real as the others, as one may appreciate if he tries to steer a car around a curve on glare ice without enough friction. However, for the present we shall discuss only acceleration along the direction of the velocity, just speeding up or slowing down along the same straight line.

In discussing an equation like $F = ma$ numerically, one needs a system of units, and there are two main systems in use. In the English system in use commercially in Britain and North America, with which most of us are familiar, the units of length are the mile, foot, and inch. In the metric system in commercial use in the rest of the world and used by science almost universally, the corresponding units are the kilometer (or 10^3 meters since "kilo" means a thousand), the meter, and the centimeter (10^{-2} meter), abbreviated km, m, and cm. We see that this is a decimal system in which one does not need to remember arbitrary numbers like 5280 and 12. A meter is about 40 inches (a little over a yard, more precisely 39.37 inches) and a centimeter is about four-tenths of an inch.

There are also metric units of mass, the gram and the kilogram, which is a thousand grams. In English units forces are measured in pounds. We are familiar with a pound as the force it takes to support a pound of butter held in the hand, for example—the *weight* of that much matter. A kilogram (or kg) is defined as the mass of a cube of water ten centimeters on a side (a liter or roughly a quart of water weighing roughly two pounds at the rate "a pint weighs a pound"). More exactly, a kilogram of matter weighs 2.2 pounds here on earth.

A convenient way to think about force is in terms of a spring balance, an old-fashioned instrument that used to be hung from a beam of a store. From a hook at its lower end was suspended a pan in which vegetables or nails could be weighed. The spring balance consists of a spring with a pointer sliding along a scale. A force applied to the hook stretches the spring, and the pointer indicates with how many pounds of force the hook is pulling.

Fig. 1–2 Accelerating a cart of mass m by means of a force pulling it.

As a numerical example of $F = ma$, we might think of using a spring balance to accelerate a small cart, pulling just hard enough to keep the pointer at a steady reading so as to accelerate the cart with a constant force F, as in Fig. 1–2. Let the mass of the cart be two kilograms and let us speed it up from being at rest to a speed of one meter per second in one second. The change of velocity is one meter/sec and it happens in one second. Then its acceleration, or rate of change of velocity, is one meter per second per second, abbreviated 1 meter/sec/sec or 1 meter/sec^2. Then $F = ma$ says

$$F = (2 \text{ kg}) \times (1 \text{ meter/sec}^2) = 2 \text{ kg meter/sec}^2.$$

This slightly awkward expression kg meter/sec^2 is a perfectly good standard or unit in terms of which to measure force. In scientific parlance it even has a special name, the newton. It arises from thinking of forces in terms of the accelerations that they cause. To compare it with more familiar things, one kg meter/sec^2 is a force equal to almost a quarter of a pound, so the spring balance, if calibrated in pounds, would have to read almost half a pound in our example.

This business about units is a necessity if one wants to be definite, but the most important thing about an equation like $F = ma$ is that it is an expression of the orderliness of nature. Despite the apparent chaos about us, there are many types of things that don't just happen but occur in a beautifully regular and predictable manner. There are relationships between events. Much of science is concerned with trying to understand and make use of these relationships.

GRAVITATIONAL FORCE

We all know that massive objects when unsupported near the surface of the earth fall toward the center of the earth. The mysterious force of gravity acts between all massive objects. The earth, being a very massive object, attracts the objects near it quite strongly, with the familiar force that we all know and live with. We observe that an unsupported chunk of metal is accelerated toward the center of the earth. We have said that it takes a force to cause an acceleration. Thus we know that gravity is exerting a force on the chunk of metal. When we hold it in our hand, we can feel the force. We know that we have to push upward on it to prevent it from being accelerated downward. To hold it still, we exert with the hand a force upward on it equal and opposite to the force of gravity, so that the total force acting on it is zero. Then $F = ma$ applies, with $F = 0$ and $a = 0$.

It is instructive to demonstrate falling bodies by dropping a metal ball and a feather at the same time inside a vacuum. They are suspended in a vertical glass tube near the top and practically all the air is pumped out with a vacuum pump. Then they are dropped and fall side by side, the feather accelerated just as fast as the metal ball. Thus, with air resistance removed, all bodies near the surface of the earth are accelerated at the same rate. This acceleration due to gravity is

$$g = 9.8 \text{ meter/sec}^2.$$

The force of gravity on any object is the force that will cause this much acceleration, and the equation $F = ma$, or in this case $F = mg$, tells the amount of the force. For example, the force of gravity on a 2-kg mass is 2 kg $\times$ 9.8 meters/sec^2 = 19.6 kg meter/sec^2.

WORK

Energy is defined in terms of work. The word work in science has a slightly different connotation from the word as we use it in ordinary life. I may push hard against a wall and feel that I am working hard. But in fact I am accomplishing nothing, and in the scientific use of the term I am doing no work. *When a force is applied to an object, work is done by the force only when the object moves and has at least some of its motion along the direction of the force.* If a car is carrying a man uphill at a steady rate, the car exerts an upward force on the man equal to his weight. Suppose that while the car goes forward 50 feet, it gains 5 feet of altitude. The work done by the car on the man is equal to the man's weight multiplied by 5 feet. This is the force applied multiplied by the distance moved upward in the direction of the force. It is the same amount of work as if an elevator lifted him vertically 5 feet. If the automobile were instead moving at a constant velocity along a horizontal road, the upward force exerted on the man by the car would be doing no work because there would be no upward motion. It would be merely supporting him at a constant level. Carrying an object at a constant speed horizontally is like pushing against a wall: there is no motion in the direction of the force and no work done. The object might as well be rolling on a cart with nothing pulling it.

In general, then, the work done on an object by a force applied to it is equal to the force multiplied by the distance moved in the direction of the force:

(work done) = (force applied) × (distance moved in direction of force).

POTENTIAL ENERGY

The energy possessed by an object is its capability of doing work. If a man lifts a ten-kilogram mass from the floor and sets it on the table, he has done work on the mass. It is now in a position to do work on something else. It could be hung by a string wrapped around a pulley and made to drive a machine while it falls back to the floor, as in Fig. 1–3. It thus has the capability of doing work on the machine. This capability that it has because of its position is known as its potential energy. More precisely, because it is the force of gravity that pulls the mass back to the floor, the potential energy may be called gravitational potential energy. In falling back to the floor the mass can do the same

Machine on which m can
do work while falling

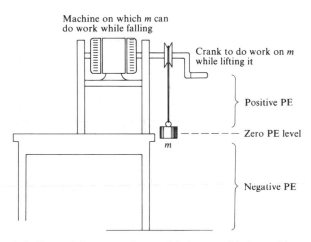

Crank to do work on m
while lifting it

Positive PE

Zero PE level

m

Negative PE

Fig. 1–3 Potential energy of mass M above and below table top.

amount of work that was done on it in lifting it to the table, namely, the weight of the mass (or force of gravity on it) multiplied by the height of the table.

If we limit ourselves to the room while discussing motions, we may say that the potential energy of the mass is zero when on the floor. Its potential energy when on the table is then equal to its weight times the height of the table.

Thus potential energy is defined in the same terms as work: it is the work done on the object in moving it from a position of zero potential energy to where it is. Potential energy is thus a relative thing—relative to an arbitrarily defined zero. The important quantity is the change of potential energy in moving from one point to another. If we should be doing experiments on the table and define the table top as the zero of potential energy, then the object has negative potential energy when it is placed below the level of the table top. The object can do work on something else while moving toward the floor, which means that a negative amount of work is done on it, so to speak. When the potential energy is negative, the magnitude of the potential energy is the amount of work that must be done on the object to get it back up to the zero level.

The concept of potential energy is put to good practical use in the large pumped storage reservoirs used in connection with nuclear and other electric generating plants. These generating stations, particularly

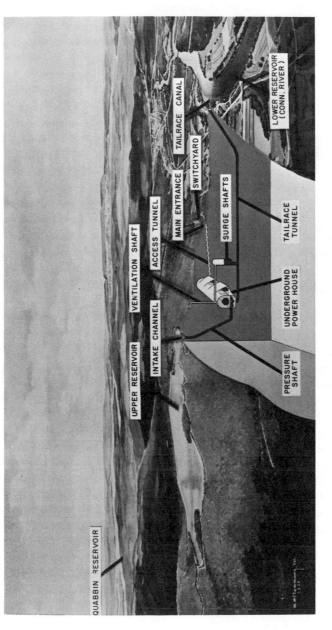

Fig. 1-4 Pumped storage installation at Northfield, Massachusetts.

Fig. 1–5 Power house excavated inside mountain at Northfield, showing turbo-generators under construction.

the nuclear ones, achieve top efficiency when they produce electric power continuously. The demand for electric power, however, fluctuates rapidly, being heaviest in the evening hours and very heavy during warm spells because of air conditioning. The idea of a pumped storage station is, then, that during slack periods the generating plant can produce electricity that is used to pump water up to a high level, converting the electric energy into potential energy of the water in a high reservoir. During peak demand periods the water descending from its high level is used as a source of hydroelectric power, driving turbogenerators to convert the potential energy into electric power when it is needed in this form. Actually, the machinery is reversible and can be used both ways, first as an electric-driven pump to raise the water and then as a turbogenerator driven by the water to produce electric power. It is important to appreciate that this is not a way of producing energy but only of storing energy that has already been produced as electric energy, a way of "putting it on the shelf" for future use.

One such reservoir is proposed at Storm King Mountain on the Hudson River near New York City, and another is being constructed in a mountain on the Connecticut River at Northfield, Massachusetts, in connection with the nuclear generating plant being built at Vernon, Vermont. At Northfield the turbogenerators are in a large room deep within the granite of the mountain (see Figs. 1–4 and 1–5). Nuclear reactors could be installed underground similarly if the expense could be justified for the sake of added safety.

The pumped storage reservoir illustrates very well the meaning of potential energy; the water at high elevation has the potentiality of doing work as it comes down to river level. In actual practice so much energy is lost in the pumps and pipes and generating machinery that only about 60 percent is recovered as useful electric power. Still the fuel wasted is cheap enough that it pays to store it this way to supply the peak demand.

Very substantial amounts of energy may be stored in this way. As an order-of-magnitude estimate, suppose such a reservoir were 100 meters wide by 1 kilometer long (about the area of 20 football fields) and 10 meters deep, making a volume of 10^6 cubic meters or a weight of 10^9 kg, since a kilogram is established as the weight of $(0.1 \text{ meter})^3$ of water (or one liter or 10^{-3} cubic meters of water). Suppose the mean height of the reservoir is 200 meters above river level. The downward gravitational force on each kilogram of water as it descends through that height is 9.8 kg meter/sec^2 (or roughly 10 kg meter/sec^2), and the work done on it by gravity is 2000 kg meter^2/sec^2. Thus the total potential energy stored in the reservoir is 2000 × 10^9 or 2 × 10^{12} kg meter^2/sec^2. Electric energy is measured in kilowatt hours and the conversion factor

$$1 \text{ kwh} = 3.7 \times 10^5 \text{ kg meter}^2/\text{sec}^2$$

leads to the result that the potential energy of the water in the reservoir is equivalent to about 5 × 10^5 kilowatt hours or 500 megawatt hours. A typical nuclear generating plant, such as the one at Vernon, Vermont, may have a generating capacity of about 500 megawatts, so such a reservoir could store its output for about an hour. Since the peak demand lasts only a few hours, this is the order of magnitude of the storage capacity needed.

Potential energy under the influence of the force of gravity is not the only type of potential energy. Elastically distorted bodies have potential energy associated with their distorted shape or with the position of their parts. The stretched sling shot, the hunter's bent bow, the compressed spring all are capable of doing work as they return to their undistorted position. When distorted, they have a potential energy equal to this work they can do or to the work that was required to distort them.

KINETIC ENERGY

A moving object, even if it has no force at all acting upon it, is capable of doing work by virtue of the fact that it is moving. A good example is the energy of a hammer being used to pound a nail in a wall, just before it hits the nail. Because it is a moving mass, it is capable of doing work on the nail. This capability of doing work by virtue of its motion is called kinetic energy. It is equal to the amount of work that was done on the body in speeding it up from rest. It is also equal to the amount of work that it can do while being slowed down to rest. It is, for example, the average force that it will exert on the nail times the distance the nail is driven. The force exerted by the hand that accelerated the hammer was a weaker force acting through a greater distance.

When a body is dropped so that it falls freely under the force of gravity, its potential energy is converted to kinetic energy as it accelerates. The force of gravity is working on the object to create its kinetic energy as the potential energy decreases. This is an example of the conservation of energy: if there is no friction to dissipate energy, then the total energy of the falling object is constant. Potential energy (PE) decreases as kinetic energy (KE) increases.

Another example is given by a horizontal force F applied to a massive sled m sliding on a frictionless horizontal plane. Here the sled is supported by a vertical force on the runners exactly equal and opposite to the force that gravity exerts downward on the sled (that is, its weight) and the vector sum of these two vertical forces is zero. Thus although there are three forces acting on the sled, the total force is equal to the horizontal force and this is what accelerates the sled.

As long as the horizontal force F is applied, the sled accelerates. When the force is no longer applied, the sled coasts along at a steady velocity, v. By virtue of its velocity it has kinetic energy, equal to the amount of work done accelerating it. This means that it has the capability of doing work as it decelerates, perhaps by pulling on a string that is wrapped around the shaft of a machine or perhaps by smashing something with which it collides.

The amount of kinetic energy of such a moving body is given by the formula

$$KE = \tfrac{1}{2} mv^2.$$

Here the factor m is easily understandable; for a given velocity, twice as much mass means twice as much kinetic energy. Two equal objects moving along side by side have twice as much kinetic energy as one of them, and their total kinetic energy is not changed if they are joined together to make one object. But we see that the formula contains not just v but v^2. Not only does the kinetic energy increase with increasing velocity but, for a given m, it increases so rapidly that twice the velocity means four times the kinetic energy.

This is to be understood by thinking of how much work it takes to accelerate the body. When the force is along the line of motion, we remember that the work done is equal to the force applied multiplied by the distance the object moves while the force is applied. Let us compare the work done during the first part of the acceleration with that during the last part of the acceleration. Suppose the force is great enough to make an acceleration of $\frac{1}{10}v$ per second so that it takes ten seconds to speed the object up to velocity v. During the first second, when the object speeds up from zero to $\frac{1}{10}v$, it is moving very slowly (with an average speed $\frac{1}{20}v$) and moves only a short distance, so the work done during that first second is small. During the last second the same amount of force speeds the object up from $\frac{9}{10}v$ to v at an average speed of $0.95v$, much faster than before, so it moves much farther in that last second. Again, the work is the force times the distance so is much greater during that last second. It is much harder to accelerate it after it is already going fast, not because it takes more force but because it takes more work. A horse accelerating a sled moving fast has to run very hard to keep ahead of the sled and still exert the same force on it that is quite easy to exert when it is just starting. So if we add a little velocity when the velocity is already great, we add more kinetic energy than we do by adding that same velocity when the velocity is small. That is the way it is with v^2 in the formula rather than just v: the jump 9 from 4^2 to 5^2 is larger than the jump 5 from 2^2 to 3^2, for example.

For the sake of readers who like a bit of algebra, the same idea is presented again in Appendix 2 a little more fully, with some equations.

CONSERVATION OF ENERGY

The main reason we make so much of the concept of energy in discussions of natural phenomena is that it is a conserved quantity. Under a wide variety of circumstances the total energy remains constant while it is being transferred from one form to another. The transfer of energy from one form to another is what concerns us in the generation of electric power. We have found a simple example of the conservation of energy in the transfer of potential energy into kinetic energy as a stone falls. If it is a dense stone or a lump of lead, so that air resistance is practically negligible, the gain in kinetic energy is equal to the loss of potential energy and the total energy of the lump is constant. If you crumple up a piece of paper into a wad and drop it alongside the lump of lead, you will observe that the wad of paper does not gain speed as rapidly and falls behind. Thus the kinetic energy of the wad of paper is not increasing proportionately to its loss of potential energy and the total energy of the wad of paper itself is not constant because it is losing energy to the air through air resistance. The lost energy appears as heat energy in the air, or kinetic energy of the air molecules. In this case we have to look at the wad of paper and air together as the complete mechanical system, and the total energy of the system is constant. Potential energy of the paper is being transferred into kinetic energy of the paper and of the air molecules.

Another example is the swinging of a pendulum, such as in a grandfather's clock, Fig. 1–6. At the bottom of the swing we may call the *PE* zero—that is as low as the pendulum bob can go. There the energy is all *KE*. At the highest point of the swing, just as the bob is stopping and about to start back, there is no *KE* and the energy is all *PE*. On the way down, *PE* is being transformed into *KE*, and on the way back up the other side this *KE* is being transformed back into *PE*. Aside from small losses due to air resistance and other friction, the sum of the two is constant, a nice illustration of the conservation of energy as it changes from one form to another.

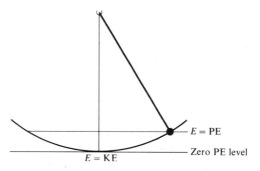

Fig. 1–6 Swinging pendulum with constant energy, $E = PE + KE$.

ENERGY AND POWER

If one does an act just once, like lifting a weight from the floor to the table, it requires a certain amount of energy, in this case the weight times the height. A continuous lifting process, on the other hand, requires energy to be expended in a continuous fashion. Pumping a stream of water up through that height requires a certain amount of energy per second, and that is called a certain *power*. Power is the rate of doing work, or the time-rate of expending energy.

An electric generating plant is rated in terms of kilowatts. A big one generates a million kilowatts, the rate at which the plant can turn out electric energy. The kilowatt, or kw, is a unit of power. To deliver a kilowatt for an hour takes 60 times as much energy (burning 60 times as much coal, for example) as to deliver that kilowatt for a minute. In our electric bills we pay for the energy delivered, or the power multiplied by the time we used the power. That is measured in kilowatts multiplied by a unit of time, in the energy unit kilowatt-hour, or kwh. We pay a few cents for a kwh to light a one-tenth-kilowatt light bulb for 10 hours.

REVIEW QUESTIONS AND PROBLEMS

1. A 1-kilogram mass and a 2-kilogram mass each have a velocity of 10 cm/sec. How does the momentum of the second compare with that of the first (that is, how many times as large is it)? How do their kinetic energies compare?

2. The 1-kg mass moves at 10 cm/sec and the 2-kg mass moves at 20 cm/sec. How do the momenta compare? Their kinetic energies? Same questions for 1 kg at 20 cm/sec and 2 kg at 10 cm/sec?

3. If a heavy stone is dropped from rest and accelerates at 9.8 m/sec^2 for 10 seconds, how fast will it be moving? If an auto is already going 10 miles an hour and accelerates at 3 miles per hour per second for a quarter of a minute, how fast will it be going?

4. If the zero of potential energy is taken to be at the level of a table top, how does the potential energy of a ball one meter above the table compare with the *PE* of the same ball 50 cm below the table top?

5. If a heavy stone falls freely from rest at the level of the table top, how does its kinetic energy at the instant it passes the level 50 cm lower compare with its potential energy at that level? What is its total energy (*KE* + *PE*) there?

6. If a cable from a high crane is lifting a half-ton of cement to the top of a building under construction, how does the power required to lift it to the top in a minute compare with the power required to lift it up there in half a minute? How does the work done in the two cases compare?

7. When a car accelerates, the power of the motor makes the tires push backward on the road, which makes the road push forward on the tires and car through friction. How does this forward force on a 1-ton car accelerating at 10 cm/sec^2 compare with that on a 2-ton car accelerating at 15 cm/sec^2?

CHAPTER 2

Some Features of a Power Plant

PRESSURE AND TEMPERATURE

The basic principles discussed in the previous chapter are at work in many ways in the nuclear reactor of a power plant. The reactor itself, a source of heat to be discussed later, is contained in a thick-walled pressure vessel of stainless steel, perhaps ten feet in diameter and thirty feet high and with walls half a foot thick to stand the enormous pressure inside. What is this pressure? Pressure is a measure of force exerted on a surface per unit area, and is commonly expressed in pounds per square inch (psi). If the reactor is of the boiling-water type, the upper part of the reactor vessel is full of high-pressure steam, the water around the reactor core being heated so hot that it boils in spite of the very high pressure. (Steam is of course the gas form of the substance that as a liquid is water. When the water boils it is converted from liquid to gas. The temperature at which boiling occurs is higher at higher pressure.) Steam consists of water molecules in violent agitation, very tiny particles each with a lot of kinetic energy (for its size). As in any gas, the hotter the steam the higher is the kinetic energy of the particles. The average kinetic energy is directly proportional to the temperature (if we measure temperature on an appropriate scale starting at the very cold "absolute zero").

These tiny particles bounce against each other, and when they hit the wall of the pressure vessel they bounce back from it. During the short time a particle is bouncing from the wall it pushes on the wall, and the wall equally pushes on it. The momentum of the particle is changed. This requires that the wall exert a force on the particle, and there is an equal and opposite force exerted on the wall by the particle.

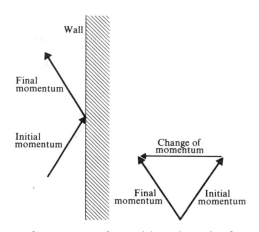

Fig. 2–1 Change of momentum of a particle on bouncing from a fixed surface.

Sometimes the particle may strike the wall squarely and bounce straight back (with the same amount of momentum in the opposite direction), in which case the change of momentum is twice the amount of momentum of the particle. A correspondingly large average force is required during the short time of the encounter, first to stop the particle and then to accelerate it back. But almost all collisions with the wall are slantwise, as in Fig. 2–1, with the particle bouncing back with the same speed it had before and making the same angle with the surface, but deflected in a different direction.

The change of momentum is then figured with the vector diagram shown, in which the original momentum, the change of momentum, and the final momentum make three sides of a triangle. This is the way vectors add, with the tail of one arrow at the point of the arrow to which it is added. The triangle shows that the original momentum plus the change of momentum equals the final momentum. It shows that the change of momentum is less than twice the magnitude of the original (or final) momentum, and would be very small at close to grazing incidence when there is only a slight deflection. Thus the average momentum transfer is considerably less than twice (and indeed about equal to) the average amount of momentum per particle. In their helter-skelter motions the particles do not all have the same speed, but we may speak of average kinetic energy or average amount of momentum.

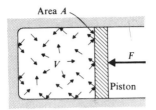

Fig. 2-2 The force to hold the piston in place, and thus the force exerted on the other side of the piston by the gas, is equal to the pressure of the gas times the area of the piston.

The relationship between pressure and force can be illustrated by thinking of a cylinder with a sliding piston as one of the walls of the volume containing a gas, as in Fig. 2-2.

The piston would recoil to the right from the force exerted on it by the bouncing particles if there were not an equal and opposite force F applied to the piston toward the left to hold it in place. We may say that the piston merely transmits the force F to the particles as they bounce from the surface, and F is equal to the total rate of change of the momentum of all these particles. This rate of change of momentum is the average change of momentum per impact multiplied by the number of impacts per second. One might expect such a force to be jittery but there are so enormously many molecules in a small volume of gas (something like 10^{20} per cubic centimeter) that the force is quite steady.

Other things being equal, the larger the area of the piston the greater the applied force F must be. This idea leads to the concept of pressure. The pressure of the gas is that force divided by the area A of the piston,

$$P = \frac{F}{A} \quad \text{or} \quad F = PA.$$

That is, pressure is force per unit area and thus the force is the pressure times the area.

If the volume is increased by moving the piston to the right while care is taken to keep the temperature and thus the mean KE constant, the gas particles have more room to move around in and fewer of them per second will hit the piston. The pressure is thus decreased by increasing the volume at constant temperature. But if volume is

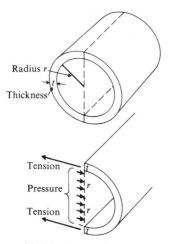

Radius r

Thickness

Tension

Pressure

Tension

Fig. 2–3 The tension provided by the tensile strength of the metal must be great enough to compensate for the pressure of the gas.

kept constant, increasing the temperature (and thus the mean KE) naturally increases the pressure. (Indeed the pressure increases proportionately to v^2, the square of the particle velocity, or to the average KE, because the faster the particles the greater the momentum change per impact *and* the greater the number of impacts per second.) The result of this is that the pressure is proportional to the temperature divided by the volume, or

$$PV = KT.$$

The numerical value of the constant K is directly proportional to the amount of gas (or number of molecules) in V.

One of the features of a nuclear reactor is the very thick wall of the pressure vessel. To get a rough idea of the thickness required to stand the pressure inside, think of a section of a steel cylinder, a part of the pressure vessel, as shown in Fig. 2–3, and think of it as divided into two halves by an imaginary plane down the middle as shown. The two halves are pushed apart with equal and opposite forces.

The total force toward the right-hand half of the cylinder, for example, is most easily calculated by noting that this force on the cylinder itself is the same as the force with which the two halves, cylinder-plus-gas, push each other apart due to the pressure along that imaginary plane. (It may help the imagination to think of a thin sheet of metal foil along that plane as a buffer between the two halves of the gas. True, the pressure pushing on the inside of the right half of the cylinder is exerted over a larger surface than that of the dividing plane but partly in the wrong directions, so the total force *toward the right* is the same as the force across the plane.) Like pressure, the tensile strength of steel is measured in pounds per square inch, the number of pounds of pull per square inch of cross section that it will tolerate without breaking (or the maximum pull of a 1 in. × 1 in. steel bar). By requiring, across that imaginary plane, that the steel be at least strong enough so that the force holding the two halves together is equal to the force of the gas pressure pushing them apart, we see that

(tensile strength) × (thickness of wall) = (pressure)

$$\times \text{ (inner radius of cylinder)}$$

or

$$\frac{\text{thickness}}{\text{radius}} = \frac{\text{pressure}}{\text{tens. strength}}.$$

The hemisphere that caps the cylinder at each end of the pressure vessel needs to be only half as thick because its surface curves in two directions.

The tensile strength of the best stainless steel is about 10^5 psi which means that a vessel of 5-foot radius and half a foot thick, which are realistic dimensions, would ideally contain a pressure of 5000 psi. Actual pressures attained in reactors are considerably less than this, about 2000 psi, because of the need for a "factor of safety." This is particularly important in nuclear reactors. One reason is that the neutron flux causes radiation damage that might weaken the steel, while corrosive action eats away the inner surface. But the most important reason for a safety factor is that the amount of radioactivity stored inside that might escape if the vessel should burst may be several hundred times as large as the radioactivity produced by an atomic bomb, as is discussed later.

As a further line of defense against the possibility of a lethal release into the atmosphere, the reactor vessel and its immediately associated machinery is enclosed in a much larger steel shell, containing air near normal atmospheric pressure so that workers can move around in it outside the radiation shield surrounding the reactor vessel. This large outer shell, much of it below ground but showing a dome-shaped top, is what one sees in the photographs of some nuclear power plants. The outer shell is in most cases designed to be strong enough to contain the pressure that would result if the high-pressure contents of the inner reactor vessel possibly superheated by a runaway accident of the reactor itself, should escape and be diluted to a much lower pressure in the larger volume of the outer shell. (Here we remember $PV = KT$.) The outer shell may be about 60 feet in diameter and 2 inches thick or it may be of thicker concrete. It has the advantage of not being subject to radiation damage, being outside the radiation shield.

Because of uncertainties about what might happen in an accidental burst, such as whether the reactor (inner) vessel might not just spring a leak but instead explode like a hand grenade with fragments piercing the outer shell, or what might happen in a bombing attack or by sabotage, some advisors have urged that reactors should be built in underground cavities for the sake of stronger containment. It has been estimated that this would increase the overall cost of a power plant by not more than about 5 percent. However, only a few reactors in Sweden and Switzerland have been built underground.

The idea of pressure is also important in the functioning of the steam turbine. The turbine converts the kinetic energy of the heated gas particles into the mechanical energy transmitted by a rotating shaft to the generator, where it is in turn converted to electric power as the useful output of the plant. In a nuclear plant or a conventional coal-burning power plant, high-pressure steam is conducted through pipes to an orifice, where it is permitted to escape into a region of lower pressure. The pressure behind the orifice accelerates the steam to a high velocity, so it escapes as a high-momentum jet, impinging on the curved blades of a rotating turbine wheel in such a way that the stream is deflected backward as the blades move forward. Just as in the case of a single particle deflected by the vessel wall, so the stream of particles deflected by a blade exerts a forward force on the blade equal to the rate of backward change of momentum of the stream, and this force drives the turbine. Actually, the turbine is cleverly designed so that the jet, after it has expanded in the lower pressure and has been deflected backward, is deflected forward again by a larger curved stationary blade. From this the steam impinges on another rotating blade at a still lower pressure. This cycle is repeated for several stages as the pressure of the gas is reduced to that of the condenser at the outlet end of the turbine. Many such jets work on many blades of each turbine wheel at the same time.

The turbine is thus designed for high efficiency, but no matter how well it is designed, there is a fundamental limit to how much mechanical energy may be derived from the steam as its temperature and pressure are reduced from that of the boiler to that of the condenser. Cooling water flowing through the condenser is heated while the steam is condensed, to become warm water that is pumped back to the boiler.

COOLING WATER AND CARNOT EFFICIENCY

The most important output from a nuclear plant is of course the electric power carried away by transmission lines. Another output, undesirable but inevitable, is the heat that passes out through the cooling water. This either heats the river or other body of water, possibly to the detriment of its ecology, or is dissipated into the atmosphere by means of cooling towers, usually along with great quantities of moisture (Fig. 2–4). The kinetic energy of the molecules of a gas and the pressure they can exert is proportional to the temperature. The steam that goes into the turbine at high pressure and high temperature comes out at a lower pressure and lower temperature. If there were no way to conduct away the heat, the temperature and pressure would rise at the output end until ultimately they would be as high as at the input end and the turbine would stop. Even when there is cooling by river water, the kinetic energy and pressure of the gas at the outlet end cause a back pressure reducing the efficiency of the turbine.

Energy goes into the turbine as high-pressure steam. Some of it is converted to electrical energy by the turbogenerator, and the remainder goes down the river as lower-temperature heat. For this reason there is a theoretical limit to the efficiency of utilization of the power from the boiler, even if the machinery were as perfect as possible. This limitation is associated with the name of an early French physicist, Nicolas Carnot. It is stated in terms of the absolute zero of temperature, the temperature at which the kinetic energy of agitation of atoms and molecules would be reduced to zero. We should think of the temperature scale as given in Fig. 2–5. In the United States most people

Fig. 2–4 The Calder Hall Nuclear Power Station in England, with its enormous cooling towers and their plumes of water vapor.

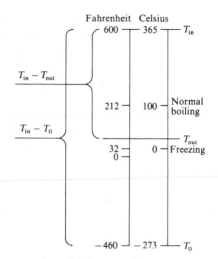

Fig. 2–5 Temperature scales, with input and output temperatures of a machine, and absolute zero.

are familiar with the Fahrenheit scale in which water normally freezes at 32 degrees and normally boils at 212 degrees and we talk of summer temperatures of 90 degrees. On this scale reactors boil water at around 600 degrees and the absolute zero is −460 degrees. The centigrade scale, at which water normally boils at 100 degrees and freezes at 0 degrees, is shown beside it. Absolute zero is −273 degrees centigrade.

The steam goes into the turbine at T_{in} and goes out to the condenser and cooling water at temperature T_{out}. The bottom of the scale, absolute zero, we call T_0. The theoretical maximum of energy that can be converted to useful work, or the ideal efficiency, is now

$$\text{Efficiency} = \frac{\text{useful energy}}{\text{input energy}} = \frac{T_{in} - T_{out}}{T_{in} - T_0}.$$

Since T_{out} is more or less determined by the nature of the cooling water available, a fixed amount of energy corresponding to the lower end of the scale below this is wasted. Thus efficiency is improved by making T_{in} as high as possible. For a given output power, increasing the efficiency in this way reduces the amount of heat wasted to heat the river or atmosphere. Actual efficiencies of nuclear plants, somewhat less than the theoretically possible, run around 30 or 32 percent. Those of large modern coal-fired plants such as are now being constructed (with higher allowable boiler temperatures) may exceed 40 percent and may heat the river only two-thirds as much as nuclear plants do for a given electric output. Thus conversion to nuclear power tends to increase what is sometimes called the thermal pollution of a river, but there is some of it in either case.

The temperature of steam is a measure of how much *KE* there is per pound of water molecules. The molecules in liquid water or in solid water (ice) also have energy of thermal agitation proportional to the temperature. When water boils, the molecules have such violent thermal agitation that many of them jump out of the liquid surface and into the gas (steam). They do this both at the top surface of the water in a teakettle and at the surface of bubbles forming near the bottom where the heat is being applied.

The rate at which molecules in the steam hit the surface and return to the liquid depends on the pressure. If the pressure is high enough, more molecules return to the liquid than evaporate away and the boiling stops. At high enough pressure for a given temperature bubbles do not form and the liquid does not boil. The boiling temperature or "boiling point" increases with pressure as shown in Fig. 2–6. If the pressure

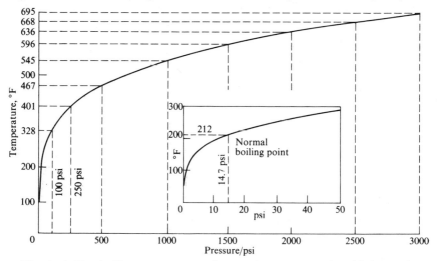

Fig. 2–6 The boiling temperature of water increases greatly with increasing pressure.

is kept constant by letting steam escape just fast enough, the temperature does not keep on going up beyond the boiling point as heat is applied. Instead the energy is spent forcing the molecules away from the water surface and thus changing water into steam at constant temperature. Thus in a boiler in a power plant, steam is normally formed at a nearly constant pressure and temperature.

In early electric generating plants the steam was used to drive a reciprocating steam engine like that in a steam locomotive. Such an engine has a sliding piston in a cylinder something like the piston-cylinder system in an automobile engine. As the piston moves away from the high-pressure steam pushing it, the molecules bounce back from the receding surface with reduced speed compared to the speed they would have if the piston were not moving. Imagine bouncing a tennis ball against the back of a truck moving forward, compared to bouncing it against a stationary truck. In the first case the ball doesn't bounce back as fast.

The gas molecules thus lose kinetic energy while they do work on the piston, and the temperature becomes lower as the gas expands. In a modern steam-electric plant the steam emerges in a jet pushing against the rotating blades of a turbine and similarly does work on the receding blades as the molecules bounce back with reduced kinetic energy.

ELECTRICITY

One cannot understand nuclear power generation without having some idea of what electricity is, for nuclear energy starts with an electric push between two parts of a fissioning nucleus as they fly apart and ends as electric current in a power line. Between these extremes the energy changes many times from one form to another.

There are several types of forces between material bodies. Gravity is one such force, a force of attraction. Between very massive bodies— the earth and moon, for example—it is a large force. But it is a force that acts between any two masses. Between two bodies small enough to handle indoors it still exists, but is so weak that we do not ordinarily notice. There are other forces known as electric forces, acting between certain special types of bodies that are electrically charged.

The classic demonstration of a charged body is a glass rod that has been rubbed with silk or a hard-rubber rod that has been rubbed with fur. We say the glass rods have a positive electric charge, and the rubber rods have a negative electric charge. We also say that like charges repel while unlike charges attract one another. This can be shown by suspending one of the rods by a long silk thread and bringing the other close to it. If the rods are alike they will move apart; if one is rubber and the other glass they will move toward each other.

By rubbing more or less vigorously, one may make the charge and the consequent force larger or smaller. One can then ask the philosophical question: Is there a smallest possible charge, or can we go on dividing the charge into smaller and smaller pieces? Clever experiments investigating very small charges in special conditions under a microscope have shown that there is, indeed, a smallest possible charge. One type of tiny particle that carries the smallest possible charge is an electron. All electrons are the same, each having a definite mass (about 10^{-27} grams) and a definite amount of negative electric charge.

Individual electrons are too small to see, so it may be hard to get some feeling for their reality. One comes fairly close to seeing electrons when watching a television picture. The space behind the screen is a vacuum, practically completely empty of gas such as the air we breathe. This makes it easy for electrons to pass across that space when shot into it. The light of the picture is made by electrons hitting the back of the screen, a stream of electrons cleverly timed and steered so as to form a coherent picture.

Wires of copper and similar metals have the property that electrons can flow through them quite easily, and they contain enormous numbers of electrons ready to flow when pushed. Electrons flowing through a wire constitute an electric current and the power is carried away from a power station essentially by having an excess number of electrons at the power station end push other electrons along the power line, the push being passed on from one group of electrons to another. There is a close relationship between electricity and magnetism, and at the power station the electrons are pushed by the rapidly changing magnetic conditions in the electric generator (and nearby transformers).

The wire also contains positive charges, the positive nuclei of atoms, that do not move about. There is everywhere in the wire about as much positive as negative charge so its net charge is zero, even while it is carrying a current of electrons pushing each other through it. Glass and rubber differ from the wire in that electrons do not flow easily through them. They are insulators, such as are used to support power lines to prevent the current from flowing to the earth. Rubbing them in the demonstration just mentioned removes or adds some electrons to leave them with a net positive or negative charge.

Even in metals, electrons do not flow with complete ease, and in some metals there is more resistance to their flow than in others. Because of this resistance, energy is lost in heating a wire when a current flows through it. More energy is lost in some kinds of wire than in others, and more is lost in a thin copper wire than in a thick copper wire for a given amount of current. Copper and silver have relatively little resistance and are considered very good conductors of electricity. Aluminum is also quite good.

The current, which is proportional to the number of electrons passing per second, is measured in terms of a unit known as the ampere. Because of resistance, it takes a certain amount of electrical "push" to make an ampere go through a given segment of wire in a closed circuit of wire. That electric push, or potential difference, is measured in volts, so it is also known as voltage. A bright electric light bulb is ordinarily designed for an electric potential difference of about 100 volts, which pushes through it a current of about 1 ampere. The electric power (the amount of energy per second) required to do this is measured in watts. In this case the power is about 100 watts, the number of volts multiplied by the number of amperes or the voltage times the current. The more usual commercial unit is the kilowatt, kw, 1000 watts.

Another natural force akin to gravitational force and electric force is the force of magnetism. When a magnetic compass needle is placed near a wire carrying current, the needle is turned in a certain direction. A current exerts a force on a nearby magnet and can cause it to move. Conversely, a magnet moving past a wire that is part of a closed loop can cause a current to flow in the wire. In more detail, the moving magnet causes a voltage in the nearby part of the wire, and this drives a current around the closed circuit.

An electric motor works by having currents push magnets, thus turning shafts to do mechanical work. An electric generator develops a voltage to drive a current by having magnets move past wires, and it requires external work (such as from a steam turbine) to keep the magnets moving. In both, the magnets are maintained by currents passing around pieces of iron, the iron serving to enhance the magnetic effects of the currents.

This whole process is reminiscent of the exchange between *KE* and *PE* in a simple pendulum. Energy is continually being transformed from one form to the other. Similarly, the motor and generator between them can transform electric energy into mechanical energy and back again. Indeed, a motor-generator combination is sometimes used to transform a large current at low voltage into a small current at high volta·e (or vice versa) by way of mechanical power transmitted by a shaft in between. The total power or number of watts is slightly smaller for the output than for the input because of small heat losses in the machine.

If converted in an ideal generator without heat loss, $\frac{3}{4}$ foot-pound per second of mechanical energy produces 1 watt of electrical energy, or 1 horsepower produces $\frac{3}{4}$ kilowatt. These and other such conversion factors are tabulated in Appendix 13.

A current in a wire wrapped around a piece of iron makes a magnet of the iron, a stronger or weaker magnet depending on the strength of the current. A stationary magnet of varying strength is just as effective as a moving magnet in inducing a current in a wire. Thus a current that is continually varying in strength in one wire can induce a varying current in another wire by its varying magnetic effect around it. The effect is particularly strong when both wires are wrapped around an iron or steel core in a device known as a transformer.

Alternating currents which continually vary their flow from one direction to the other have particular uses in power transmission. Steady or direct current can be stepped up or down in voltage by use of a motor-generator that is expensive. Alternating current can be stepped up or down in voltage with a transformer that is much cheaper and easier to maintain. (Also, alternating current is easier to shut off than direct current in case of line failure.)

Stepping the voltage up and down is important because high voltage is dangerous and inconvenient in the home or factory or generating plant but is economical for long-distance transmission. The reason for this economy is that resistance losses are proportional to the current flowing, not to the voltage. At high voltage less current is required to transmit a given amount of power through wires of a given thickness, and hence there is less resistance loss. Near a power plant one sees a group of large transformers surmounted by a framework to hold the emerging high-voltage lines far apart, supported on long insulators so sparks will not jump between them. From there the familiar big towers to support the well-separated lines extend across the countryside. The voltage may be about a quarter of a million volts, to be stepped down through several transformers by a factor of 2000 before reaching our light bulbs.

REVIEW QUESTIONS AND PROBLEMS

1. If the cylindrical part of the main pressure vessel in a reactor is 6 feet in diameter and 6 inches thick, while cylindrical walls of the outer containment vessel are 30 feet in diameter and 2 inches thick, how do the maximum contained pressures for the two compare?

2. If the diameter of the piston of a large car is twice the diameter of a piston of a small car, how does the force on the two pistons compare when each is subjected to a pressure in the cylinder of 100 psi?

3. A coal-burning power plant has a boiler temperature of 640° Fahrenheit and a condenser temperature, cooled by a lake in winter, of 40° F. A nuclear plant has a boiler temperature of 540° F and a condenser temperature, rather high because it uses cooling towers, of 90° F. How do the ideal thermal efficiencies of the two compare?

4. Does a glass rod that has been rubbed with silk attract or repel a rubber rod that has been rubbed with fur? Do two glass rods (suitably rubbed) repel or attract one another? If we could observe the effect on a single electron, would the electron be attracted or repelled by a glass rod (suitably rubbed)?

5. How would a compass needle behave near the end of a long magnet?

6. Why is alternating current, an electric current continually changing in direction, more useful than steady or direct current for power transmission?

CHAPTER 3

Atoms, Molecules, and Nuclei

THE SMALL SIZE OF ATOMS

In a conventional power plant burning "fossil fuel" (coal, oil, or gas) the energy is released from atoms (or molecules composed of atoms) and in a nuclear plant the energy is released from the nuclei of atoms, so we should know something about atoms and their nuclei.

We might be content simply to say that an atom is very small and its nucleus is very much smaller: the size of an atom is about 10^{-8} centimeters and the nucleus is about 10^{-12} centimeters. Thus a nucleus is about one-ten-thousandth as big as an atom. The dot representing the nucleus at the center of Fig. 3–1 is drawn large enough to be seen, about a hundred times smaller than the diameter of the pictured atom. On a true scale it should be about a hundred times smaller still.

To go a little farther, we may say that the atom is like a little solar

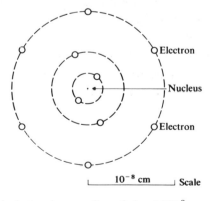

Fig. 3–1 A typical atom has a radius of about 10^{-8} cm.

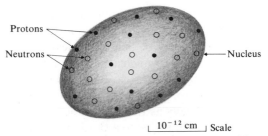

Fig. 3–2   A nucleus has a radius of about 10^{-12} cm.

system, with electrons orbiting around the central nucleus in circles, as shown in Fig. 3–1, while the nucleus at its center is in some ways like a tiny water droplet and in some ways like a capsule of gas molecules or a mad swarm of bees in a rubber balloon, as in Fig. 3–2. But these descriptions are a bit inadequate as a basis for discussing the way energy is generated in reactors, and we shall delve a little deeper into the nature of atoms and nuclei.

There are presumably two kinds of readers of this book. One reader is impatient to get on to the practical industrial applications and their social consequences and wants only the bare essentials of the physics needed as a background. The other reader takes pleasure in understanding the physics a little better and appreciates that this helps him think a bit more deeply about the technical problems involved. At this point in our travels toward nuclear energy there is a parting of the ways. The path straight ahead in the text is a shortcut. The other path is an optional detour; the reader who wishes to follow the detour and learn more about the nature of atoms should turn to Appendix 4 and return to the main text again at page 47.

ATOMS AND WHAT THEY ARE MADE OF

We have learned that an electron is a tiny particle with a discrete mass and a discrete electric charge. If a stream of electrons shot through a vacuum (for example, inside a television tube) should pass close by a rubbed glass rod that has a positive electric charge, the path of the electrons would be deflected and would bend a little bit toward the glass rod. There is another kind of tiny charged particle known as a proton. A stream of them would be bent away from the positively charged glass rod. Protons have a positive electric charge, an amount of it called e, and electrons have an equal amount of negative electric charge, $-e$. The proton is about 2000 times as heavy as the electron, and its path would accordingly be bent much less.

An atom consists of a heavy nucleus surrounded by some electrons. The nucleus contains some protons, and different nuclei contain different numbers of protons. But all nuclei (except hydrogen) are about twice or more as heavy as the sum of the protons in them.

The difference in weight is due to a third kind of particle that is an important constituent of the nucleus. This particle is called a neutron. It has about the same mass as a proton (actually about 1 percent greater), again about 2000 electron masses. But the unusual thing about the neutron is that it has no electric charge. A stream of them passing the charged glass rod would not be deflected by it and would follow a straight path.

A nucleus is made of protons and neutrons. There are many different kinds of nuclei and many different kinds of atoms. The number of protons in a nucleus is called Z, the number of neutrons is N, and the total number of nucleons, as the protons and neutrons are called, is $A = Z + N$. Electrons are so light that the mass of the atom is made up almost entirely of the nucleon masses, so A is called the atomic mass number.

The electric charge on the nucleus is Ze (the sum of the positive charges of all Z protons). The number of electrons surrounding the nucleus in a normal atom is also Z. The sum of the charges of all Z electrons is thus $-Ze$, so the total charge on the atom is zero. That is what makes it a normal atom. It can exist without attracting other charges to it from far away.

The electrons move around the nucleus in a sort of mad scramble that still has some regularity to it. Most of them stay quite close to the nucleus but a few venture farther out, being only slightly more strongly attracted by the positive nucleus than they are repelled by the near-in electrons. They race about in such a way that the atom is roughly spherical. The edge of the atom is thus not sharply defined but instead rather fuzzy. Nevertheless, one can say that some atoms are larger than others. The way the size of atoms varies with their number of electrons, Z, is shown in Fig. 3–3.

The sharp peaks in the curve are one of many indications that the electrons are arranged in an atom in rather fuzzy shells; some groups

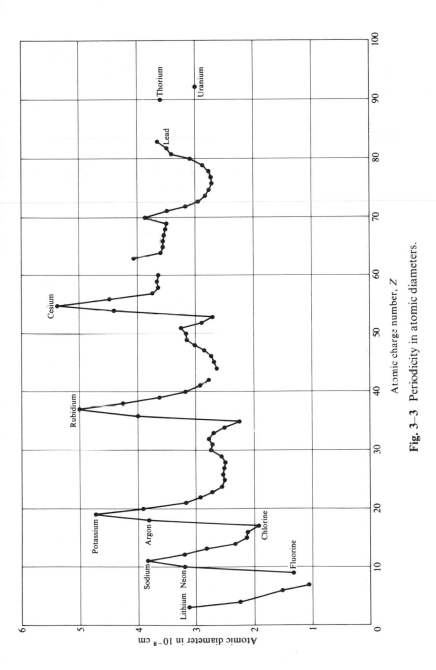

Fig. 3–3 Periodicity in atomic diameters.

of them have larger average distances from the nucleus than others do, as they race around. Only certain numbers of electrons are allowed in each shell by the natural laws of motion governing their behavior.

In the neon atom, with $Z = 10$, the shells are filled up (or "closed"), and again in the argon atom with $Z = 18$. The sodium atom, with $Z = 11$ and the potassium atom, with $Z = 19$, thus each have just one electron outside closed shells. Now, the single electron outside of closed shells is only weakly attracted to the rest of the atom because there are $Z - 1$ electrons in nearer the nucleus and pushing it out almost as hard as the Z protons in the nucleus pull it in. It thus tends to wander rather farther away than other electrons do, and an atom with a single electron outside closed shells is larger than others. This accounts for the peaks in the figure at sodium and potassium, and the other peaks are similar.

This variation in size of atoms and variation in the looseness or tightness of binding of the outer electrons to the atom determine the chemical properties of atoms. Figure 3–3 is an illustration of the way these properties vary *periodically* as the various shells are filled, so that sodium is chemically similar to potassium, fluorine is chemically similar to chlorine, etc. These latter two nuclei, with $Z = 9$ and $Z = 17$, each have one electron missing from the last shell, an electron that would be rather strongly attracted to the nucleus if it were there. These holes in the shell are hungry for that missing electron, and if they can grab the loosely bound outer electron from a sodium or potassium atom, they do so and the two atoms stick together to form a molecule such as sodium chloride, or salt.

An important conclusion from all this is that the chemical energies associated with the formation of molecules are made up of differences of the energies with which outer electrons are bound to atoms. The outer electrons of heavy atoms with large Z are not much more energetically bound than are those of light atoms with small Z because with large Z there are more inner electrons to almost cancel out the greater nuclear charge.

Both kinetic energy and potential energy are associated with the internal motions of the electrons within an atom, and the sum of the two is the internal energy of the atom. We have seen that it is quite arbitrary how we define the zero of potential energy (at the table level or floor level, for example). The forces holding the particles in an atom (or a nucleus) together become extremely weak if the particles get very far apart, and the potential energy changes very little as they move around at great distances. The potential energy of such collections of particles is usually defined to be zero when all the particles are very far apart. This means that the potential energy of an electron in an atom is negative. It has fallen in from far away, at potential energy zero, toward the nucleus that attracts it, just as a mass starting at potential energy zero at the table top falls into a position of negative potential energy near the floor, toward the earth that attracts it. Kinetic energy is always positive, but the kinetic energy of the electrons in a normal atom is smaller in magnitude than their potential energy so the total internal energy of the atom is negative. The electrons stay in the atom because they do not have enough energy to get far out to where their potential energy is zero. Electrons are thus bound in the atom by this negative energy of binding. A marble may similarly continue to roll around in a dish, not having kinetic energy enough to climb up over the edge.

Atomic energies are measured in a unit called the electron volt, or ev. The electron volt is the energy required to push a charge e through a potential difference of one volt. (A flashlight battery cell has a voltage of 1.5 volts, meaning that the amount of energy required to push one electron (in a vacuum) from one terminal to the other is 1.5 ev.) The energies with which outer electrons are bound to normal atoms vary from about 3 to 20 electron volts, 3 to 7 ev for a single electron outside closed shells and 15 to 20 ev when the outer electrons are in a closed shell or almost-closed shell. The total energy of the electron in question is negative by this amount. It takes this amount of additional energy to knock the electron out of the atom (which is like making the marble roll over the edge of the dish).

MOLECULES

Molecules are aggregates of two or more atoms sticking together by means of electric forces on electrons behaving according to the quantum rules. A simple example of molecule formation occurs when a sodium atom meets a fluorine atom, $Z = 11$ meets $Z = 9$. The sodium atom has an extra electron hanging loosely on the outside, and the fluorine atom has a hungry vacancy in its outer shell. The extra electron

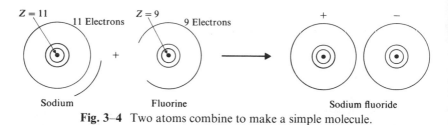

Fig. 3–4 Two atoms combine to make a simple molecule.

is attracted into the vacancy, and the energy of the whole assemblage becomes lower if the extra electron becomes a member of that outer shell of fluorine, where it can feel a stronger nuclear attraction (only partially shielded). The simplest way to describe the resulting molecule is to say that we now have a positive sodium ion and a negative fluorine ion held together by the electric attraction between their opposite charges, as in Fig. 3–4. They each have 10 electrons (with charge $-10e$) in closed shells, just like the neon atom, but the sodium has a nuclear charge $+11e$ so a net charge of e, while the fluorine ion has a nuclear charge of only $+9e$ or a net charge of $-e$. This molecule is called sodium fluoride. The more common sodium chloride, or table salt, is similar in composition.

Not all the bonds holding atoms together to make molecules (or crystals) are so simple, but they all depend on the same principle, namely, that some of the electrons find advantageous ways of arranging themselves, within the limitations of the quantum rules, to feel the pull of two nuclei rather than just one and thus to lower the energy of the assemblage. That is one reason the mechanics of the atom are so important. They are the basis for all the differences in the behavior of various atoms to make molecules, and molecules are the basis for the rich variety in chemistry that makes *us* possible, among other things.

The energies by which atoms are bound together to make molecules thus involve slight rearrangements of the electrons from the way they are bound in the single atoms, and therefore are of the same order of magnitude as (and almost always somewhat less than) the energies by which electrons are bound to atoms. In most molecules having only two atoms, the atoms are bound to one another by one or a few electron volts.

There are enormously many kinds of molecules. A certain kind of atom can exist in any one of a number of kinds of molecules. It is

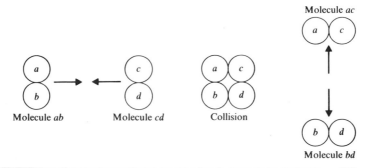

Fig. 3–5 In a chemical reaction, two molecules react to make two different molecules.

important that its energy of binding may be quite different in one molecule from that in another. This makes it possible for us to gain energy by letting the atoms be rearranged, starting with molecules in which they are relatively loosely bound and ending with molecules in which they are more tightly bound and have more negative internal energy.

Chemical Burning and Explosives

Chemical burning is what happens when coal burns in a furnace, for example. In the simplest case we can imagine two molecules, one composed of atoms $a + b$ and another of atoms $c + d$, moving quite fast and colliding with one another, as in Fig. 3–5. When the four atoms come so close together that the electrons can easily jump from one to another, the electrons find their energetically more favorable spots, pulling the four atoms even more rapidly together. Then they bounce apart in their new combinations with increased kinetic energy, a and c forming one molecule and b and d another. Commercially important cases of the burning of fuel are mostly more complicated than this, involving the simultaneous collision of three molecules rather than two, but the general idea is the same. For example, methane molecules, each with one carbon and four hydrogen atoms, may meet two oxygen molecules, each with two oxygen atoms, and they come apart with increased kinetic energy as two water molecules (one oxygen and two hydrogen each) and a heavy carbon dioxide molecule (a carbon atom with two oxygen atoms). In chemical symbols this is written:

$$CH_4 + 2O_2 \rightarrow CO_2 + 2H_2O.$$

This is the most important reaction in the burning of natural gas. In the burning of coal or oil there is also another reaction in which hydrogen and oxygen molecules combine to make water molecules. Fuel oil contains about two hydrogen atoms for each carbon atom, on the average. The energy appearing as heat when fuel oil burns yields about $2\frac{1}{2}$ electron volts per atom of fuel. For coal the figure is about half that. These figures, arising from differences of energies of various molecules, are thus seen to be comparable to but somewhat less than the energies by which the electrons are bound in atoms.

An explosive is a device that releases a lot of energy in a small space in a short time. The sudden expansion of suddenly heated material in air causes a shock wave to travel outward and carry much of the released energy in a way that can be very destructive. The destructive power of a chemical explosive comes not because an exceptionally great energy is released per kilogram, but because it is released suddenly. In fact, chemical explosives release only about as much energy per kilogram as does the burning of coal, but coal burns much more slowly and the energy is carried away gently as heat. Both the explosives and the coal get their energy release by trading off the binding energies of the outer electrons of atoms, and these amount to only a few electron volts per atom, as we have seen. The burning of coal requires the relatively slow circulation of air to supply the oxygen that combines with the carbon and hydrogen of the coal. Chemical explosives contain all the ingredients needed; all that is required to cause the reaction that releases energy is an impulse to push the molecules together.

Some explosives detonate more suddenly than others. Gunpowder is relatively slow. It is fast enough to build up a high pressure in a small space in a gun barrel before the projectile has moved much, but not so fast as to damage the barrel. Explosives like TNT that are known as "high explosives" detonate locally much more suddenly. They are a solid waxlike or plastic material, and the detonation that starts at one spot builds up a sudden pressure that passes on to detonate the neighboring material. In this way a powerful shock wave of increasing strength passes across the explosive material and consumes it.

A shock wave is the passage of a very precipitous high-pressure front. Ahead of it the pressure remains normal until extremely suddenly the pressure becomes very high and then gradually diminishes as the shock wave passes. The pressure profile is shown in Fig. A3–1 of

Appendix 3, where shock waves, sound waves, and other waves are further described.

A shock wave can also travel in air and other media as a result of an explosion, but with gradually diminishing rather than increasing intensity while propagating as an expanding sphere. Pressure variations (such as sound waves) travel faster at higher pressure; this means that all the little pressure variations in the high-pressure region behind the shock front travel fast enough to catch up to the shock front and add to its intensity. They cannot get ahead of it because the velocity is low in the low-pressure region ahead of the wave. That is why the front is so sharp. Meeting an obstacle, a shock wave delivers a sharp blow and can be very damaging.

LIQUIDS

The strong forces holding atoms together to make molecules are known as valence forces. They derive their strength from the fact that the electrons take advantage of the shell structure of the atoms in the formation of the molecules, as we have seen. Even after all this advantage provided by the shell structure has been used up to form the molecules, there are still some weak and very short-range forces acting *between* molecules. When two molecules come very close together, the electron shells of the atoms in each molecule can be slightly distorted in a way to reduce the total energy of the pair of molecules, and the two molecules then tend to stick together. The energies involved here are something like one-tenth of an electron volt or less, for each pair of neighboring molecules. Each molecule feels an interaction of this sort with each of its nearest neighbors.

In a liquid the molecules cluster about one another, practically touching each other, held together by these weak short-range forces. At low temperatures they stick rigidly together in a solid mass, each molecule continuing to hold onto its same neighbors in the same shape around it. At somewhat higher energies the solid becomes a liquid, with the molecules vibrating so violently that they easily slip past one another and the liquid can flow. As they slip around, neighbors may interchange but each atom remains in quite close contact with all its neighbors. The volume of a large mass of molecules is thus the sum of the volumes of the individual molecules, and the liquid can only be

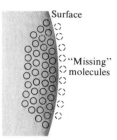

Fig. 3–6 Surface tension. The molecules on the surface of a liquid are attracted by fewer neighbors than are those inside.

compressed a very little bit under pressure. Unlike a gas, a liquid thus has a constant mass per unit volume, or constant "density."

Each of the molecules in a liquid has something like ten neighbors. Each pair of neighboring molecules gives rise to a negative potential energy corresponding to an attractive force. Since the number of molecules per unit volume remains constant, there is a definite amount of binding energy per unit volume.

The surface of a liquid is particularly interesting. A molecule on the surface obviously has fewer immediate neighbors than a molecule in the interior, as suggested in Fig. 3–6. As compared with a molecule in the interior, there is some binding energy missing that makes the total energy of this volume of liquid higher than it would be if it were part of a continuous distribution of liquid. Surface "costs" binding energy, so to speak.

Complicated systems like this when allowed to "settle down" tend to get into a shape or position having as low energy as possible. A sample of liquid having no other forces acting on it takes on a spherical shape because a sphere has the least surface area for a given volume. If there were a few molecules in a bump on the surface, they would be pulled in by the forces toward the rest of the liquid and squeeze in to take their place in the sphere. The surface thus behaves very much as it would if it were a stretched rubber membrane. Hence the term "surface tension," which is measured as an energy per unit of surface area.

NUCLEI

The nucleus of the atom, tiny as it is, has a very important structure of its own. Atomic dimensions are measured in 10^{-8} cm. Nuclear dimensions are measured in 10^{-12} cm or 10^{-13} cm. The nucleus is like a drop of liquid in one important respect: it has constant density, and all nuclei have about the same average density, as though they were made of the same kind of "liquid" or "nuclear fluid."

Measuring the Size of Nuclei

The apparently constant density of nuclei was discovered by measuring the sizes of various nuclei, considering them to be like little droplets not far from spherical in shape. This is done by throwing something at a nucleus and seeing how it is deflected, taking an average for many tries. The projectile used is either a proton or a bare helium nucleus, called an alpha particle (containing two protons and two neutrons tightly bound together). The trick of throwing a stream of protons at a target full of nuclei (within their atoms in a gas or solid) is an interesting one and involves an elaborate machine known as an accelerator. There are various kinds of accelerators. Some of them accelerate the proton along a straight line—linear accelerators—and some around a circular path between the poles of a magnet—cyclotrons, for example.

In principle, the simplest type of linear accelerator consists of a proton source and a long tube "full of vacuum," from which practically all air and other gas has been pumped out, with a strong positive charge at one end and a strong negative charge at the other end so the protons receive a strong electric push as they pass from one end to the other. They pass through a hole in the end carrying the negative charge and through more evacuated tubes as they hit the target and are observed to bounce off in various directions from it.

The proton source, at the positive end, consists of a space containing hydrogen gas under fairly low pressure and an "electron gun" to shoot electrons at the hydrogen atoms and knock the electrons off them, leaving bare protons to be pushed by electric charges toward a small hole where they enter the long vacuum tube. The "electron gun" consists of a hot wire (like that in a light bulb) from which electrons are thrown out, by their high kinetic energy at high temperature, into the low-pressure space containing hydrogen gas, where they are accelerated by electric charges in the general direction of the hydrogen atoms. When the electrons are knocked off, the bare protons find their way through the little hole and on down the path of the "beam." The hole

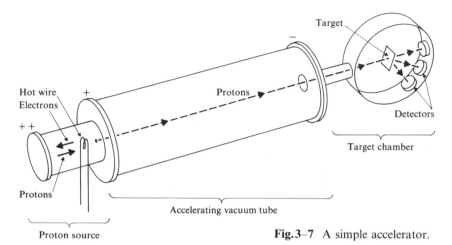

Fig.3–7 A simple accelerator.

is so small that good pumps can take care of the little hydrogen gas that leaks through and maintain a good enough vacuum. Sounds like Rube Goldberg, and looks like that in Fig. 3–7, but it works. Real accelerators in big laboratories look a lot more impressive (Fig. 3–8), but that is the principle of the thing. There are other types of accelerators, such as the cyclotron, that use different principles to supply similar beams of protons or other ions.

Now let us think what goes on in the target, on a different scale. There are many target nuclei, but think of just one as in Fig. 3–9(a). If the proton misses it by a lot, the proton goes practically straight forward, but if it is a near miss it feels the strong electric repulsion between the two positive electric charges (the positive nucleus and the positive proton) and it is deflected sideways. This force is strong when the charges are close together, but becomes rapidly weaker as they are farther apart (according to an inverse-square law that makes them one-fourth as strong when they are twice as far apart, for example. This is discussed in Appendix 4). The proton may have a large kinetic energy of several million electron volts (abbreviated Mev) and it has to be a very near miss indeed to be deflected very much. There are more far misses than near misses so most of the protons go very nearly forward. The electrons in the target are so light that the energetic proton can push some of them out of its way and hardly notices them as it goes by.

Fig. 3–8 A van de Graaff accelerator, with outer pressure tank and shielding hoops, and dome removed to show the proton source and accelerating tube.

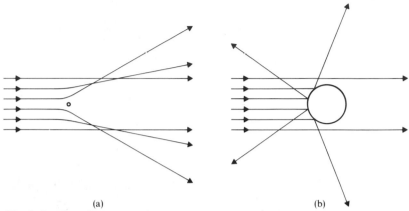

(a)   (b)

Fig. 3–9 There is more backward scattering if there is energy enough to hit the target.

The situation would be very different if we were shooting with a pea-shooter at a billiard ball (Fig. 3–9b). Then all the peas would go exactly straight except those that actually hit. Of those that hit, some just graze past the edge for a small deflection and more hit rather squarely to be deflected far sideways or backward.

By observing how the protons bounce off, one can tell whether some of them are getting in close enough to hit the target nuclei. Whether they get in there in spite of the electric repulsion depends on the radius of the nuclei. This is how nuclear radii are measured.

The result of such measurements, comparing light and heavy nuclei, is that *the volume of a nucleus is directly proportional to the number of nucleons it contains.* In symbols, the radius R of a nucleus containing A nucleons is given by

$$R^3 = r_o^3 A$$

where r_o is the measured constant, a sort of comparison radius, $r_o = 1.3 \times 10^{-13}$ cm. This applies for A varying from 4 for helium to about 240 for uranium and on beyond for some of the new man-made nuclei.

Nuclear Forces and the Nuclear Fluid

The fact that there is direct proportionality between the volume of a nucleus and the number of nucleons it contains means that it makes sense to talk about a sort of "nuclear fluid" having constant density.

However, the reasons for this constant density are very different from the reasons for constant density in an ordinary liquid. A liquid has constant density because its constituents, the molecules, have a definite volume and are quite closely packed together. In the nuclear fluid the nucleons of which it is made are very much smaller than the volume they occupy and they fly around past each other in somewhat the same way as the atoms in a gas or the electrons in an atom.

The constant density is associated with the nature of the forces between nucleons. These are a very special type of force, not due to electric charges on the nucleons, and they depend on the distance between the particles in a very different way from the electric forces between electric charges that hold atoms together. The potential-energy curve representing the *electric* forces has a long "tail" gradually fading away at great distances. Far out the potential is very slightly negative, and the curve bends gradually downward as the distance is decreased (the *potential* becoming twice as deep at half the distance as shown in Fig. A4–5—a bit different from the inverse-square law for the *force*).

The potential energy curve representing the peculiar nuclear interaction between two nucleons instead remains zero until they get quite close together (about 2×10^{-12} cm) and then plunges down quite suddenly to a deep negative potential energy at distances shorter than about 1.5×10^{-13} cm, as sketched in Fig. 3–10. There r_{ab} means the distance between nucleon a and nucleon b. It is a deep and short-range

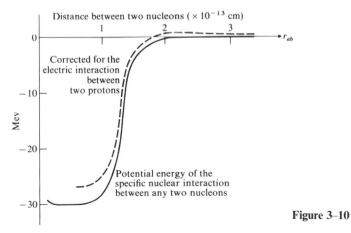

Figure 3–10

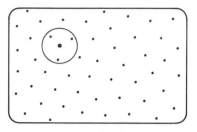

Figure 3–11

potential energy curve. The interaction has another peculiarity that can be described in simplified terms by saying that one nucleon can interact with no more than three other nucleons at a time when they come within range. If a fourth comes within range while three are already that close, the fourth just doesn't feel any force of attraction.

With a curve like that, the *PE* between two nucleons is low if they are about 10^{-13} cm apart, but it doesn't get any lower if they are closer. Imagine a lot of nucleons flying about helter-skelter in a box of volume V, as in Fig. 3–11. As long as the box keeps them close enough together that on the average there are almost always three nucleons within the potential energy range of each nucleon, then the *PE* will not be much reduced if V is reduced to squeeze them together more, but it will be increased sharply if V is increased to let them get out of range much of the time. The system seeks to get in the situation of lowest total energy, *KE* plus *PE*. If we squeeze the nucleons together more than enough to make the *PE* low, the influence of the *KE* takes over. As those who have followed the detour through Appendix 3 have seen, the quantum rules that govern the behavior of these tiny particles make the *KE* increase as the volume becomes small. This incipient increase of *KE* prevents the nucleons from squeezing together more than is required to make the *PE* about as low as possible. The compromise of these two influences keeps the nucleons about as far apart as the range of forces, on the average. The *PE* and accompanying force pulls in, but not all the way, and the *KE* tends to disperse the nucleons as far as the *PE* will easily allow. This is a way to account approximately for the constant density.

Now let us think about the potential energy of a single nucleon as it interacts with others while swimming about in that nucleon soup. The number of other nucleons within range of it may fluctuate a bit,

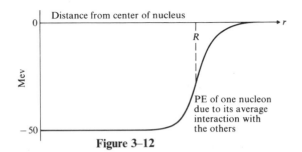

Figure 3–12

but its average *PE* does not depend on where it is as long as it is in the interior of the volume. If it gets too near the edge or just outside the edge of where the others are, it has fewer nucleons in range, and its *PE* shoots up as it feels a pull back in toward the others. Here we have the basis for understanding why a sort of "droplet" of nuclear fluid has a surface tension trying to hold it in a spherical shape, much the way a normal liquid droplet has.

If we think of the other nucleons as forming a spherical droplet, then the curve for PE_{av}, the average potential energy of the one nucleon, may be sketched as in Fig. 3–12. This looks very much like Fig. 3–10 but has a different meaning. It is again a deep "potential well" but with the "range" now given by r, the radius of the nucleus.

So this is the shape of the potential that each nucleon feels as it moves about in the nucleus. It is different from the potential in an atom such as $-e^2/r$ in the hydrogen atom, being flat-bottomed rather than having a very deep hole in the middle. This marked difference between the atom and the nucleus arises from the existence of an all-powerful central object at the center of the atom exerting the preponderant force, around which all the others revolve, whereas there is no such preponderant object in a nucleus. It is the difference between a dictatorship and an ideal democracy. In the atom, there is a big boss at the center dictating the behavior of all the other individuals. In the more democratic nucleus, all the individuals have equal influence in determining how the others behave. The *PE* curve in Fig. 3–12 represents the attraction that one nucleon feels due to the presence of the other nucleons around it. The flat bottom means that a nucleon can run around inside the nucleus quite freely, changing the nucleons with which it interacts as it goes, but when it comes to the edge where the potential curve rises, it bounces. The nucleon behaves something like a ball rolling in a flat-bottomed bowl of which the *PE* curve is a cross section.

Here we come to another "detour" sign. At this point the reader who wants more development of this subject should turn to Appendix 10 to learn how the nucleons in a nucleus behave to make it what it is, how they are arranged in "shells" that are in some ways similar to those of electrons in an atom and in some ways different. Then he should return here.

We have seen in Fig. 3–3 that atoms with the same amount of filling of their outer shells have like properties as a result of shells filled with 10 electrons (neon) and 18 electrons (argon), for example. Both have closed outer shells. The corresponding shell-closing numbers in nuclei are 8 and 20. The complete list as we go on to heavier nuclei is 2, 8, 20, 50, 82, 126, and (theoretically) 184. The quantum rules determining how many nucleons can go into a shell apply *separately* to neutrons and protons. Oxygen-16, with 8 neutrons and 8 protons, is an example of a very stable closed-shell nucleus. The alpha particle, or helium-4, with 2 of each, is another.

There are two quite different ways of describing a nucleus, both of which are approximately correct. In one view it is like a drop of fluid. Within the volume of the fluid it has approximately constant binding energy per nucleon, but at the surface some of this binding energy is lost. In a more detailed view each of the nucleons has a characteristic state of motion determined by the quantum rule and has a corresponding energy level. This is part of the machinery that makes the nuclear fluid behave the way it does, so to speak.

An atom, with each of its electrons influenced mainly by the big-boss nucleus at the center, remains practically spherical whether or not it has a closed-shell number of electrons. The more democratic nuclei are more pliable in this regard, subject to mass movements. Closed shells tend to make them spherical and stable. An example is lead-208, with one closed shell of 82 protons and another of 126 neutrons. Nuclei with almost-closed shells, for example having neutron numbers within about 5 of being equal to 82 or 126, are also spherical. Nuclei that do not have almost-closed shells are more flexible. For them the droplet idea is even more meaningful than for spherical nuclei, for they are like droplets deformed into an ellipsoidal shape. It is significant that uranium and plutonium are of this type, with proton and neutron numbers far from those closing shells. Their deformability contributes to the way they can be "split" to release nuclear energy.

For these nuclei the shell idea is also of importance in describing their tendency to be elongated. The extra nucleons not in closed shells tend to bunch together at opposite ends of the next unfilled shell. Or we may say that within the droplet the nucleons can spread out and thus reduce their kinetic energy without increasing the volume if as many of them as permitted by the quantum rules run back and forth in the same direction and stretch out the nucleus in that direction.

With protons and neutrons as the building blocks, nature has done a remarkable job of making a great variety of ways they can stick together to make nuclei—about 300 or so ways in all. Each type of nucleus is defined by its number of protons, Z, and of neutrons, N. These are plotted in Fig. 3–13, where each dot represents one kind of nucleus. One is struck by the way they cluster together in a long "region of stability" that starts out for light nuclei with N equal to or only slightly greater than Z, but curves upward on the chart to make N very considerably greater than Z for heavy nuclei.

Notation for Nuclei

The standard notation used to describe succinctly these various nuclei, giving their proton number Z, neutron number N, and nucleon number $A = Z + N$ along with a reminder of the chemical name belonging to Z, is

$$\begin{array}{c} A \\ Z \end{array} \text{Ch}_N$$

Here *Ch* indicates where the abbreviated chemical name of the element Z should stand. By looking at the little dots in Fig. 3–13, we see that there exist five stable nuclei each having $Z = 20$. Each of these is normally the center of an atom with 20 electrons, an atom of the chemical element calcium. These are the five stable "isotopes" of calcium. They range from neutron number $N = 20$ to $N = 28$. The heaviest of them is thus Ca-48 or ${}^{48}_{20}\text{Ca}_{28}$. They all behave the same chemically, since chemical reactions are determined by the numbers of electrons filling the electron shells, which in turn are determined by the number of protons. Thus Z identifies the chemical elements.

It is precisely because the isotopes behave alike chemically that it

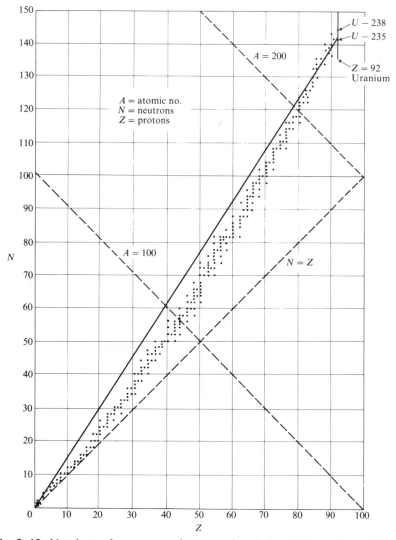

Fig. 3–13 Numbers of neutrons and protons for all the stable nuclei and the almost-stable alpha emitters beyond proton number 82.

is so hard to separate them in the nuclear-energy enterprise. Some elements have only one stable isotope, but most have two or more. Other, unstable, isotopes can be made in nuclear reactions. Uranium, the heaviest natural element, has two isotopes so nearly stable that they exist in nature, U-235 and U-238 or

$$^{235}_{92}U_{143} \qquad \text{and} \qquad ^{238}_{92}U_{146}$$

Notice again that these values of Z and N are far from the respective shell-closing numbers 82 and 126.

NUCLEAR BINDING ENERGIES

One very remarkable property of nuclei is that the binding energy per nucleon is almost the same (about 8 Mev) in all but the very lightest nuclei. This again is closely analogous to the situation in an ordinary liquid, where the amount of heat energy required to boil a liquid away (that is, to separate it into its constituent molecules) is proportional to the amount of liquid boiled away. Binding energy, we remember, is the amount of energy required to take the system apart and put all the particles far away from each other. Conversely, it is the energy that could be gained by letting the particles come together in an act of fusion. As has been stressed before, the total energy E of such a bound system is negative and the binding energy BE is a positive quantity, the magnitude of that negative energy.

In nuclei the binding energy situation is very different from that in atoms, as shown in Table 3–1, with examples helium, lithium, and carbon for atoms; helium, strontium, and uranium for nuclei.

TABLE 3–1
Binding energy of atoms and nuclei

Atom	Z	BE/Z	Nucleus	Z	N	A	BE/A
He	2	19 ev	$^4_2He_2$	2	2	4	7.0 Mev
Li	3	68 ev	$^{88}_{38}Sr_{50}$	38	50	88	8.7 Mev
C	6	145 ev	$^{238}_{92}U_{146}$	92	146	238	7.7 Mev

In this table we see a sharp contrast between the way the binding energy per electron increases so very rapidly even in the first few atoms and the way in which over a wide range of nuclei the binding energy per nucleon varies by only about 20 percent in the neighborhood of the value 8 Mev.

The very light nuclei with $A = 10$ or less contain so few nucleons

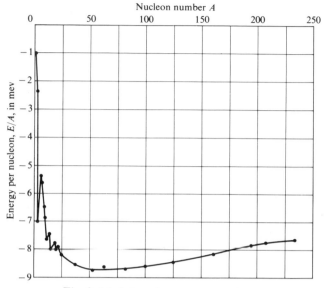

Fig. 3–14 Internal energy per nucleon.

that the volume is not so well defined, the liquid analogy does not make much sense, and the energy per nucleon fluctuates considerably. Beyond that, the curve of energy per nucleon, as shown in Fig. 3–14, is a fairly smooth curve not far from -8 Mev. However, it is very important for our purposes that it does not quite remain at a constant level. It sinks to a minimum in the region around $A = 50$ to 130 and after that gradually rises.

The energy per nucleon is thus lower in the middle of the curve than at the right-hand end. If a uranium nucleus at $A = 235$ were divided into two nuclei, having $A = 100$ and $A = 135$, for example, the curve shows that the energy per nucleon would become lower than it is for $A = 235$. Since the total number of nucleons is not changed, this means that the total internal energy is reduced and the nucleus could give off extra energy and do external work while being divided and settling down into a condition of lower internal energy. This is approximately what happens in the fission process and the curvature of this curve is the source of the nuclear energy we shall be discussing later on.

The general nature of this curve can be understood as a result of striking a balance between five different contributions to the energy. Thus the whole thing is rather complicated, but each of the contributions by itself can be explained in fairly simple terms. These five contributions to the energy are:

1. Volume energy
2. Surface energy
3. Electric energy (for simple case $N = Z$)
4. Reduction of electric energy by changing protons into neutrons
5. Increase of nucleon-shell energy by changing protons into neutrons

The first three of these are the most important, for we can understand the source of the energy of fission and fusion in terms of them in a general way. But nuclear energy would not be available from fission without the fission neutrons that have been changed from protons, and we need terms (4) and (5) to understand their origin.

Volume Energy

As in the case of the liquid we have discussed, a collection of nucleons in a nucleus has an energy which is larger for large volumes of nuclear matter than for small volumes in direct proportion to the volume—or would be if there were no surface and electric effects. The volume is also proportional to the number of nucleons, so this "volume energy" is proportional to the number of nucleons, or in other words the volume energy per nucleon is constant. This liquid analogy does not make sense for the nuclei containing only a very few nucleons, but may be applied for A greater than about 20.

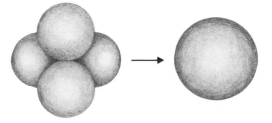

Fig. 3–15 There is less surface for the same volume when the four droplets unite.

Surface Energy

In proportion to its volume, a small nucleus has more surface than a large nucleus does. For example, in Fig. 3–15 consider a few small droplets which coalesce into one larger droplet, of course having a same total volume. The larger droplet has a smaller total surface because some of the inside surfaces have disappeared. Another way to say this is that volume is proportional to r^3 whereas surface is proportional to r^2 so that the ratio (surface)/(volume) is proportional to $1/r$.

If this statement seems unfamiliar, think of the corresponding statement for cubes. Take eight cubical blocks and build with them a big cube with twice the edge length (or twice the linear size) of one of these cubes. Some surface is lost inside. Each face of the big cube is four times as big as the face of the small cube (being made of four of them). The big cube has linear size twice as big, surface four (or 2^2) times as big and volume eight (or 2^3) times as big as the little cube. It seems reasonable that there is a similar relation between the radii, surfaces, and volumes of two spheres. This corresponds to the fact that surfaces

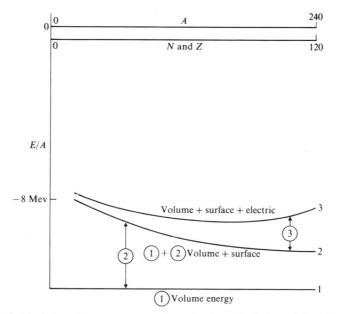

Fig. 3–16 Internal energy per nucleon for hypothetical nuclei with the same number of protons and neutrons. The volume energy, surface energy, and electric energy combine to make nuclei of intermediate mass most stable.

are measured in square feet and volumes in cubic feet, for example. Again, the surface-to-volume ratio is proportional to one divided by the linear size.

The surface energy is proportional to the surface area. This larger surface/volume ratio for small nuclei means that the surface energy per nucleon, contribution (2), is greater for light nuclei than for heavy nuclei. When plotted against nucleon number A, it is a downward-sloping curve that slopes downward more steeply for small A than for large A. This is shown in Fig. 3–16. There the bottom line, curve 1, represents contribution (1) alone, the volume energy per nucleon, E_{vol}/A, which is constant for all A and therefore a horizontal line. Curve 2 is obtained by adding contribution (2) to this, and shows the downward slope of $E_{surface}/A$.

Electric Energy

The third contribution is the electric repulsion between the protons. So far we have included only the new type of attractive force, the short-range and very powerful specific nuclear force which acts equally between all pairs of nucleons. There are three kinds of pairs of nucleons: proton-proton, proton-neutron, and neutron-neutron. The short-range specific nuclear force is the same for all three. In addition there are the more gradually varying and longer-range electric forces that are not the same for all three kinds of pairs. The neutrons have no electric charge and do not feel electric forces. Each proton has a positive electric charge and two positive charges push each other apart. There is a corresponding positive potential energy representing the work done to bring two protons together to a certain separation distance. (This again is half as great when the separation is doubled, and thus dwindles with a long "tail" at large distances, meaning that it is a long-range interaction.) This acts between all pairs of protons, and we may think of determining the total electric energy in terms of the work

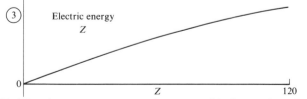

Fig. 3–17   Electric energy per proton increases with the number of protons.

required to bring in all the protons, one by one, against this repulsion. The first few brought in feel the push of only a few protons already there and are easy to bring in. The last few feel the push of almost Z protons and require more energy to get in. Thus the curve of the energy per proton must slope upward when plotted against Z, representing more energy per proton for higher Z, as in Fig. 3–17. This is contribution (3). In Fig. 3–16 we imagine the curves to be drawn for the simple case $Z = N$, as indicated by the scales at the bottom, and this contribution (3) added to curve 2 gives curve 3.

These three terms combining to make curve 3 now give us the basis for understanding the important dip in the middle of Fig. 3–14 on which the production of nuclear energy depends. When a heavy nucleus at the right end of the curve undergoes fission to make two nuclei in the dip near the middle of the curve, the two fragments have lower internal energy per nucleon (and therefore lower total internal energy) than the parent nucleus had because the electric energy is lower. This is true in spite of the fact that the two daughter nuclei together have more surface than the parent had. The electric energy is lower, one may say, because the protons are on the average farther apart after the fissile nucleus has split in two. In pushing the two fragments apart, the electric repulsion has converted internal energy into external energy. Thus the first three terms give us the main part of the story so far as energy production is concerned. The remaining terms make a small correction to this but are important for neutron production and the radioactivity of the fission products.

Neutron Excess and Beta Decay

The fourth and fifth contributions (reduction of electric energy and increase of nucleon-shell energy by changing protons into neutrons) may seem strange, for they depend on a remarkable trick by which neutrons can become protons or protons can become neutrons under appropriate circumstances. In this sense, nucleons are somewhat interchangeable.

The process by which this happens is known as beta decay. When an electron is ejected from a nucleon or nucleus with a lot of energy it is known as a beta particle (for historic reasons), and in quantity these electrons are known as beta rays. When a neutron becomes a proton, the electric charge $+e$ does not just appear all alone. Instead, the total charge is kept unchanged as zero by having the neutron simultaneously emit an electron as it is becoming a proton. Thus zero charge becomes $+e$ plus $-e$, still totaling zero, an instance of the conservation of electric charge:

$$\text{neutron} \rightarrow \text{proton} + \text{electron}.$$

(To be more exact, one should add " + neutrino," but that is a detail that has no practical consequences for this discussion.) This is the beta decay process. A free neutron, all alone in the world and not bound to any other nucleons, only lasts about half an hour on the average before it disappears by beta decay. A neutron in a nucleus can become a proton by beta decay if the nucleus gains binding energy thereby, and the electron flies off with the excess energy. In certain nuclei there occurs the converse transformation:

$$\text{proton} \rightarrow \text{neutron} + \text{positive electron}.$$

The positive electron that flies off with the excess energy in this "beta-plus" decay is a light particle like the ordinary electron except that it has charge $+e$ and lasts only a short time. It does not occur in nature. Beta-plus decay occurs only in nuclei. In free space the proton is the stable form of the nucleon.

So far we have discussed the energies of hypothetical nuclei with $N = Z$, but actual heavy nuclei have an excess of neutrons with N considerably greater than Z. This comes about because if a nucleus with a fixed total nucleon number A were to have $N = Z$, it could achieve lower total energy by having some of the protons become neutrons through beta-plus decay. The number of nucleons A would remain constant and the volume would not be appreciably changed, but the electric contribution sketched in Fig. 3–17 would be reduced in magnitude because there are fewer protons. (The excess energy goes mostly into accelerating the positive electrons as they leave.)

Thus contribution (4) is a reduction in the electric energy of the nucleus. The reduction is greater for heavy nuclei than for light nuclei, because with greater Z the reduction per beta-plus decay is greater and (because of this) there are more such decays from the $N = Z$ situation.

But not all protons become neutrons, and this is because of the fifth contribution, namely, the change of energy levels occupied by the nucleons. It is a matter of not putting all of your eggs in one basket—if you do, they are stacked up higher and have a higher energy than if they are divided equally between two equal baskets. We might think of the stack of energy levels of the various shells of a heavy nucleus as a basket to be filled with protons, and the same set of energy levels again as a basket to be filled with neutrons. (With eggs it is a matter of potential energy; with nucleons it is mainly kinetic energy.) If we move a few protons over into the neutron basket, they have to be lifted a little bit in energy, but as we move more over, we have to lift the last ones farther from a low level to a high level and the increase of energy per nucleon is greater (Fig. 3–18). Thus the advantage to be gained from the electric energy is enough to pay for moving only a limited number of protons over into the neutron category where they must occupy higher energy levels.

In Fig. 3–19 we see the result of modifying curve 3 by inclusion of contribution (4) which is negative, to give curve 4 and then adding contribution (5), which is positive, to give curve 5. This curve 5 is the final result that was sketched in Fig. 3–14.

The net result of all this is that almost all of the nuclei found in nature have N greater than Z. We see that the electric repulsion

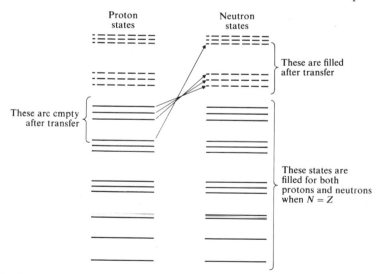

Fig. 3–18 The first shift from a proton level to a neutron level is the easiest.

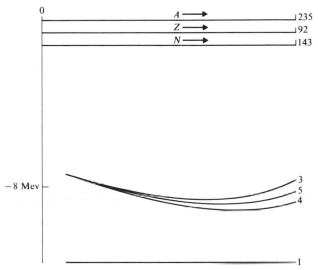

Fig. 3–19 Further changes in the energy-per-nucleon curve.

between the protons is responsible both for the upward curvature of the energy curve for high A (through contribution (3)) and for the neutron excess for heavy nuclei (through contribution (4)). In heavy nuclei each proton is repelled by so many other protons that the electric energy is very important, and N is quite a lot greater than Z. The neutron excess ($N - Z$) may be as large as 36. For light nuclei the electric effect is relatively unimportant and N is either equal to Z or only slightly larger in the stable nuclei, as we have seen in the lower left corner of Fig. 3–13.

How Nuclear Energies Are Measured: Nuclear Reactions

One way these energies E for various nuclei can be measured is by actually putting the nuclei together and observing how much energy is given off. This can be done in many steps, adding or subtracting one or a few nucleons at a time to get from one nucleus to another in a process known as a nuclear reaction.

By way of example, let us start at the beginning, building up the very lightest nuclei that will be important to us in the study of the fusion process as a possible future source of nuclear energy. These very light nuclei have names one should remember. They are shown in Table 3–2.

TABLE 3–2
The lightest nuclei

A	Z	N		Particle	Symbol	Atom
1	1	0	${}_1^1H_0$	Proton	P or H	Ordinary hydrogen, H
2	1	1	${}_1^2H_1$	Deuteron	D	Deuterium (heavy hydrogen)
3	1	2	${}_1^3H_2$	Triton	T	Tritium
4	2	2	${}_2^4He_2$	Alpha	α	Helium, He

The first three nuclear reactions with which we can illustrate the process of nucleus-building are shown in Table 3–3. In each reaction some particle is more tightly bound, so the total internal energy of the constituent nucleus is lower after the reaction than before. The total energy is conserved, and the excess energy appears in some external form, such as extra kinetic energy of the outgoing particle. The right-hand side of Table 3–2 shows the internal energies of the product nuclei deduced from the reactions, as explained later on.

TABLE 3–3
Nucleus-building reactions

Reaction	Product nucleus	Energy of product nucleus		
		A	E/Mev	E/A
1. $p + n \rightarrow D + 2.2$ Mev	D	2	-2.2	-1.1 Mev
2. $D + D \rightarrow T + p + 4$ Mev	T	3	$-2.2 - 2.2 - 4 = -8.4$	-2.4 Mev
3. $T + D \rightarrow \alpha + n + 17.6$ Mev	α	4	$-8.4 - 2.2 - 17.6 = -28$	-7 Mev

Reaction 1 is a rather unusual type, for there is no outgoing particle and the excess energy must be lost in another way. It is known as "radiative capture" of the neutron by the proton because the excess energy is lost in radiation. This is a process closely analogous to the radiation of light by atoms. In an ordinary electric discharge tube, such as a neon sign, electrons accelerated to a few electron volts of kinetic energy hit atoms having their electrons in the lowest energy levels and "excite" them to higher energy levels. In this way a hydrogen atom, for example, may be in its second energy level at $E = -3.4$ ev. In this state the atom radiates as though it were a miniature radio antenna

(electric charge moving back and forth in each case), but unlike a radio antenna it has only one amount of energy to lose as it jumps down to the only lower state at -13.6 ev. In this case the radiation is 10.2 ev worth of light, a light "quantum." One thing this light quantum could do is to meet another hydrogen atom and excite it from its lowest state to its second state, the -3.4 ev state, for it has just the right energy to do that. The second hydrogen atom is like a radio receiver, tuned to receive the radiation from the transmitting station. It is a similar radiation process in the neon atom that makes the familiar red light of a neon sign.

Various kinds of radiation are: radio waves, heat waves, light waves, x-rays, and gamma rays. These are progressively more and more energetic as we go down the list. Light waves have quanta of a few ev, x-rays of a few Kev (kilo-electron volts or thousand electron volts) and gamma rays of about one or a few Mev. X-rays penetrate through some thickness of many substances that stop ordinary light. Gamma rays are still more penetrating.

This brings us back to reaction 1. This takes place almost always when a neutron drifting along with almost no kinetic energy at all meets a proton almost at rest. Their sudden attraction to one another, when they come within range of the short-range nuclear force, accelerates the proton, and the 2.2 Mev of excess energy is radiated as a gamma ray as they "fall" into the bound state of the deuteron at $E = -2.2$ Mev. By measuring the energy of the gamma ray one knows E.

Reaction 2 is more typical of nuclear reactions. In it, an accelerator such as that described above (Fig. 3–7) is used to supply a stream of deuterons with well-known kinetic energy, KE_o, of perhaps 1 Mev or more. These impinge on a target containing deuterons, such as a thin-walled box of deuterium gas. When, in spite of the electric repulsion between them, a projectile deuteron meets a target deuteron sufficiently closely (and this is why about 1 Mev is needed), the exchange of a neutron from one to the other can take place. If a neutron jumps into the target deuteron to make it a triton, it is bound by 4 Mev more energy than it was when it was part of the projectile deuteron. This excess energy appears as kinetic energy of the outgoing proton. There is also a definite small fraction of it that goes into KE of the recoiling triton, but we shall neglect this distinction here and call it KE of the proton. That is, the KE of the outgoing proton after the reaction is $KE_o + 4$

Mev. (The term $+ KE_0$ could be written on both sides of the arrow of reaction 2 to describe it more completely, but for brevity is omitted.)

Reaction 3 is much like reaction 2, but here the captured proton, being attracted by three other nucleons, is extremely tightly bound and the neutron comes off with the large extra KE of over 17 Mev. The alpha particle is thus to be regarded as a very compact and stable nucleus, hard to knock apart. All of the nucleons are permitted in the lowest state and each attracts the other three strongly. This large energy obtainable by forming alpha particles is the main source of energy of H-bombs, and it is hoped may some day provide industrial power from controlled fusion. Of this, more later.

With the measured energies released in the three reactions, we may obtain the internal energies of the three light nuclei involved by calculating the energy balance for each reaction. This is done in Table 3–3, page 72. (The sum of the internal energies, which are negative, of the initial nuclei minus the energy released is equal to the internal energy of the final nucleus.) We see that energy per nucleon grows rapidly in magnitude for these light nuclei, with a particularly large jump from the triton to the alpha. Looking back to Fig. 3–14, one sees that these three values of E/A are plotted as the first three points at the small-A end. This, then, serves as an illustration of how that whole curve can be experimentally determined by a long succession of such nuclear reactions.

There is another and somewhat older way to determine the curve that is mentioned here mainly because of the frequent reference in popular treatises to Einstein's famous equation

$$E = mc^2.$$

This equation implies that in some fundamental sense mass is a form of energy, or that mass can be converted into energy and energy can be converted into mass. Neither alone is strictly conserved. We have spoken of the conservation of energy, which holds for everyday mechanics, but in the realm of these high-energy experiments it is only the sum of energy and its mass equivalent that is strictly conserved, the equivalence being given by this famous equation. Here E is the energy equivalent of the mass m, and c is the speed of light,

$$c = 3 \times 10^{10} \text{ cm/sec},$$

fast enough to go around the earth eight times in one second. We can see that the right-hand side must come out to have the units of an energy, for it is a mass times the square of a speed just as is an ordinary kinetic energy $\frac{1}{2}mv^2$. For example, the mass of the electron is

$$m_e = 9 \times 10^{-28} \, \text{g}$$

and if we multiply this by c^2 we obtain

$$m_e c^2 = 8.1 \times 10^{-7} \, \text{g cm}^2/\text{sec}^2.$$

When these units are worked out this quantity turns out to be about $\frac{1}{2}$ Mev, so the mass of the electron is equivalent to $\frac{1}{2}$ Mev of energy.

As an example, take the second reaction above. In an experimental apparatus known as a mass spectrograph, which consists of carefully designed magnets and plates carrying electric charges, it is possible to measure the masses of the proton, the deuteron, and the triton. Looking at reaction 2 one would think at first that the mass of the triton plus the mass of the proton would be equal to the mass of two deuterons of which they are made but this is not the case. Instead the mass measurements show that m_T plus m_P is less than $2m_D$ by about one-tenth of one percent or by about eight times the mass of an electron. This eight times the mass of an electron appears as 4 Mev in reaction 2. It actually appears as the additional kinetic energy of the outgoing proton, a direct demonstration of the conversion of mass to energy.

Shortly after 1912, when Einstein first proposed this equivalence, people began wondering whether it might be possible to "unlock the energy within the atom." The answer has come that it is possible to unlock some of it, but only a small percentage, through nuclear reactions of this sort. It was doubtless the early wording of this question that leads to the common use of the term atomic energy. Since the energy really comes from nuclear reactions, it should more properly be called nuclear energy.

FISSION

We have seen that a nucleus with the right numbers of protons and neutrons to form closed shells, and in particular the lead nucleus $^{208}_{82}\text{Pb}_{126}$ with 82 protons and 126 neutrons, is forced to be spherical by the nature of the closed shells, and that other nuclei with approximately the right number to fill shells (particularly of neutrons) are also

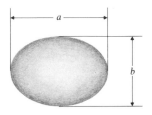

Fig. 3-20 Most nuclei with shells not almost closed are elongated.

spherical. But in nuclei having many nucleons outside of closed shells these nucleons have greater freedom of choice and find the lowest energy when they arrange their orbits in such a way as to concentrate them near one axis. The result is that such a nucleus tends to take the shape of a football, rather than a basketball, with those nucleons that are not in closed shells concentrated mostly toward the ends of the football.

Let us call the length of the football-shaped nucleus a and its thickness b, as in Fig. 3-20. If we plot the total energy, E, of the nucleus as a function of the ratio a/b, as in Fig. 3-21, the energy is lower when a is somewhat greater than b than with $a/b = 1$ corresponding to a spherical shape. This is indicated by the left-hand end of the energy curve shown in Fig. 3-21. The low point near the left-hand end of the curve corresponds to the equilibrium shape of Fig. 3-20.

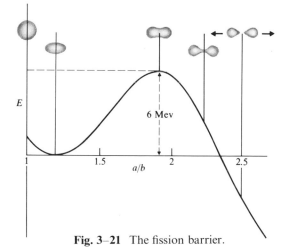

Fig. 3-21 The fission barrier.

If we think of the energy when the nucleus is stretched into a longer shape, it is convenient to consider it in terms of the liquid-drop model. The first important thing that happens is that the surface becomes larger so the surface-tension part of the energy becomes greater and the curve in Fig. 3–21 goes up. At the same time the protons that are distributed through the "liquid" are getting farther apart on the average, particularly those in the ends of the football, so their electric energy is decreasing. Nevertheless the energy goes up until the nucleus is stretched quite far. When the middle section becomes thin enough that there is only a narrow band of surface tension holding the two ends together, the electric repulsion overwhelms the surface tension and the curve turns downward toward the right-hand end. Out here where the distortion is large the two ends are pushed apart; the midsection becomes a narrow neck and then breaks to let the two fragments of the nucleus fly apart under the influence of the electric repulsion. When the fragments first separate the repulsion is very strong, and the corresponding electric potential energy is very great—about 200 Mev—because the numbers Z_1 and Z_2 of protons on the two fragments pushing each other apart are large numbers like 40 or 50 and the distance between centers of the fragments is very short, roughly the sum of two nuclear radii.

In the symbols of Appendix 4, this potential energy may be written

$$\frac{Q_1 Q_2}{r} = \frac{Z_1 Z_2 e^2}{(R_1 + R_2)}$$

where e is the charge on each proton and R_1 and R_2 are the radii of the two fragments which we approximate as two spheres in contact. This is the potential energy corresponding to a repulsive force $Q_1 Q_2 / r^2$.

As the fragments push each other apart, this 200 Mev of potential energy is converted into 200 Mev of kinetic energy of the fragments. This is the main source of the energy release in nuclear reactors and A-bombs.

The picture is then this: the short-range nuclear forces are very strong and bind the nucleons together tightly. They can bind them even more energetically in two fragments than in the big uranium nucleus where they have to compete with more protons pushing each other apart electrically.   As long as the short-range forces manage to hold the uranium nucleus together, the electric repulsion does not have a chance to push very far, but once they let go it can push the two fragments over a long distance and accelerate them to high kinetic energy. As an analogy we might think of having an auto stalled (perhaps with a dead battery) in a sharp little depression at the top of a long hill. If we can only get the car up over the edge of the little depression (with its short-range force) we can get it going fast rolling down the long hill.

We have seen that chemical reactions yield energies of the order of magnitude of 1 ev per atom involved.   Fission of the nucleus of a uranium atom yields about 200 million times as much. This is a measure of the technical revolution introduced by the discovery of fission in 1938.   It is perhaps some measure of the changes in political thinking needed to cope with it, which have not yet come to pass.

The uranium isotopes ^{235}U and ^{238}U that we find in uranium ores have been sitting for billions of years in the valley at the left-hand side of Fig. 3–21, securely held in by the surface tension and without enough energy to vibrate and become long enough to go over the fission barrier, as the hump in the curve is called.

The secret of success of the fission process is that it can be induced by absorption of a neutron. In reaction 1 above, a neutron came up to a proton and joined it to make a deuteron while radiating the excess energy away. A neutron can similarly creep up to a ^{235}U nucleus and join it to make ^{236}U. The curve of Fig. 3–21 applies to this compound nucleus ^{236}U. As we have learned in connection with Fig. 3–14, the binding energy of a nucleon to an average nucleus is roughly 6 or 8 Mev. When the compound nucleus is first formed it is in a highly excited energy state, up at the level of the broken line in Fig. 3–21, because the binding energy of the neutron permits it still to get rid of extra energy. It may do this in two ways, either by emitting gamma radiation, as in the case of reaction 1 above, or by setting the whole nucleus into violent vibration.   This energy of vibration can cause the nucleus to stretch out momentarily into a long enough shape to go over the fission barrier and proceed with the fission process. The binding energy of a neutron to ^{235}U is great enough that even an extremely slow neutron with no kinetic energy can cause fission. In ^{238}U this is not the case; the neutron must come in with additional kinetic energy of about 1 Mev or more to cause fission. A slow neutron is in general very much more easily captured by a nucleus than is a fast neutron that tends to fly on by. Thus it is much easier to cause fission in ^{235}U than in ^{238}U and the former is the more valuable isotope. Unfortunately, from the standpoint of producing power, but perhaps fortunately for the interim stability of world politics, the more useful isotope is much more rare: uranium in nature occurs as 0.7 percent ^{235}U and 99.3 percent ^{238}U.

SEPARATION OF ISOTOPES

We now come to the question of separating the two isotopes of uranium from one another or at least obtaining uranium that is richer in the rare isotope than the uranium occurring in nature. The main point here is to appreciate the enormous difference between the ease of separating elements in chemical reactions and the difficulty of separating isotopes of the same element that all behave the same chemically because they have the same number of electrons. Because of their different number of electrons, most atoms of different chemical elements are easily separated from one another, many of them in large-scale industrial processes in which are prepared many of the substances used in everyday life.

The separation of isotopes is much more costly and difficult. However, separation is less difficult for light elements than for heavy elements. The usual separation processes depend on the fact that a light molecule moves about more quickly and easily than a heavy molecule. When two kinds of molecules are mixed together in a gas at a given temperature, they have the same average kinetic energy, $\frac{1}{2}mv^2$ which means that the ones with the larger m have smaller v.

Ordinary water consists of molecules containing oxygen and hydrogen, H_2O, but not all of the hydrogen is ordinary hydrogen. There is also a small percentage of molecules containing deuterium, D_2O, that are heavier. Water molecules are taken apart to form hydrogen gas and oxygen gas by passing an electric current through water. Under the influence of the current, the H_2O molecules move faster than the D_2O molecules because of their smaller mass, and the water that is left behind has a higher than normal proportion of D_2O in it. Many repetitions of this process can concentrate the deuterium and leave very little ordinary hydrogen. This process goes relatively fast because deuterium is twice as heavy as ordinary hydrogen.

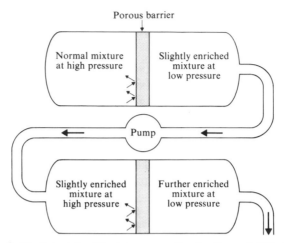

Fig. 3–22 Principle of isotope separation by thermal diffusion.

With uranium the problem is much more difficult because the two isotopes differ in mass by only a little over 1 percent, about 238 nucleon masses as compared with about 235. In this case the difficult separation is carried out in an enormous and very expensive gaseous diffusion plant that also uses a great deal of electric power. The uranium is introduced in the form of gas molecules (in combination with fluorine), and the gas is permitted to seep slowly through the very fine pores of a spongy metallic barrier (Fig. 3–22). The passages are so fine that the gas does not flow in the ordinary way but rather the individual molecules bounce between the walls of the passages on the way through and the faster ones hit the entrances of the passages more frequently and get through more quickly. This process is repeated many many times with great fans blowing the gas about through the various access chambers. Finally, from a large amount of gas at the input end a small amount of concentrated uranium enriched in ^{235}U comes out at the other end.

The first of these great plants was built in Oak Ridge, Tennessee, during World War II and others have been built since in this country (at Paducah, Kentucky, and Portsmouth, Ohio) as well as in France and China. The fact that these plants are extremely costly and sophisticated has been an important obstacle to the easy dissemination of nuclear bomb-making capability to various nations.

As a commentary on the way government institutions can work, it is interesting that the diffusion plant at Oak Ridge was built in the period 1942–44 with almost half a billion dollars that was appropriated by Congress without Congress knowing anything about it or about the highly secret atomic bomb project of which it was part. The whole atomic project involved expenditure of two billion dollars during the war and was hidden in the even larger general military budget under the distracting code name "Manhattan District" of the United States Army Corps of Engineers.

A second method for separating uranium isotopes is the electromagnetic method. It consists in accelerating molecules (positively charged molecular ions) and sending them between the poles of a magnet. This requires using gases at extremely low pressures which makes it difficult to handle large amounts of material. For this reason, this method is usually used only for further concentration or complete separation of the concentrated output from the gaseous diffusion plant.

An alternative process which has not been developed until recently is the centrifuge. Its operation depends upon the fact that a gas (or liquid) rapidly rotated in a container feels an outward force called centrifugal force, the reaction to the force required to keep it moving in a circle. Under this force the heavy molecules seep through a barrier more readily than lighter molecules. This method was abandoned in this country in the initial push during World War II, and interest in it has only recently been revived, apparently mainly in West Germany. It is said to hold promise of providing enriched uranium more cheaply and with less expenditure of power than can a gaseous diffusion plant.

REVIEW QUESTIONS AND PROBLEMS

1. If an atom has a diameter of 3×10^{-8} cm and its nucleus has a diameter of 6×10^{-13} cm, what fraction of the volume of the atom is occupied by the nucleus?

2. If a crystal is made up of atoms (perhaps two kinds of them) that average 3×10^{-8} cm in diameter and are stacked tightly together in a cubical array, about how many of them would there be in a cubic centimeter?

3. What is the special characteristic of the atoms at the peaks of the

curve in Fig. 3–3 showing the diameters of atoms of various charge numbers Z?

4. Which is greater in magnitude (that is, disregarding sign), the average potential energy or the average kinetic energy of an electron in an atom? Of a neutron in a nucleus?

5. What becomes of the atomic electrons in the formation of a sodium chloride molecule?

6. What is the essential difference in composition that makes an explosive like gunpowder behave so differently from a fuel like coal?

7. The internal energy of a drop of liquid is greater if its surface is not spherical. Why is this so?

8. How could one in principle find out how big a nucleus is, if we assume it to be nearly spherical? How does the volume of a nucleus with nucleon number $A = 200$ compare with one with $A = 25$?

9. What are the principal characteristics that a nucleus has in common with a water droplet?

10. Roughly how does the total internal energy of a nucleus with $A = 200$ compare with that of a nucleus with $A = 100$?

11. What feature of the interaction between nucleons is crucial in making the energy per nucleon less negative for $A = 230$ than for $A = 115$?

12. When fission is used to produce energy, where does the fissioning nucleus get the energy to take it over the fission barrier?

13. In a plot of nuclei according to neutron number N and proton number Z, in what direction does the line slope representing those nuclei with the same number of nucleons A? How can a nucleus change from one position to another along this line?

14. On that same plot, in what direction is a line representing emission of a neutron? Where do delayed neutrons come from?

CHAPTER 4

Nuclear Reactors
as a Power Source

CHAIN REACTION

The first property of the fission process of practical importance is that it produces two very energetic fragments, one with a little less than half and the other with a little more than half of the mass of the original nucleus. The second feature of practical importance for the release of nuclear power is the fact that these two fragments emit neutrons as they fly apart. On the average there are released about two and one-half fast neutrons per fission process. That is, each fragment emits on the average somewhat more than one neutron. The scenario of a chain reaction is, then, that an initial neutron (that either appears "out of thin air" or may be supplied by an appropriate weak neutron source) is absorbed by a uranium nucleus to induce a fission process which produces more neutrons, at least one of which produces another fission process, and so on, ad infinitum. In any practical arrangement some neutrons will be lost either by being absorbed by other substances or by escaping from the boundaries of the apparatus. It is therefore important that there be decidedly more than one neutron per fission, so as to have some to spare and still leave at least one on the average to continue the chain reaction. One principal objective in the design of an apparatus for establishing a chain reaction, either in a reactor or in a bomb, must be to minimize losses and make efficient use of the neutrons.

There are many possible examples of a chain reaction. The population explosion is one: people produce babies who eventually produce more people and so on. Only if the number of babies per couple surviving to reproduce is greater than two can the population explode.

Fig. 4–1 Model of a uranium-235 nucleus ready to release energy and emit neutrons.

A more graphic example can be made of mouse traps. A mouse trap has a spring gate that claps over, releasing energy, when a trigger is touched. If we set two corks on the gate while the trap is set as in Fig. 4–1, it will serve as a model of the fission process. A stimulus such as a cork (in place of a neutron) falling on the trigger activates the energy release and two corks (representing two emitted neutrons) fly into the air.

Imagine now setting out a number of these prepared mouse traps, each at the center of a square in a large checkerboard array, as in Fig. 4–2. The center of a flying cork must come down close to the trigger, within a small area around it, if at least an edge of the cork is to spring the trap. This effective target area around the trigger then represents the cross section of a nucleus for fission. A large array of these traps set out in their adjacent squares represents a large mass of uranium, each trap representing a uranium nucleus ready to undergo fission. We now toss in one cork to represent a stray neutron. It springs one trap which throws up two corks which probably spring one or two other traps which throw up more corks and so the chain reaction proceeds.

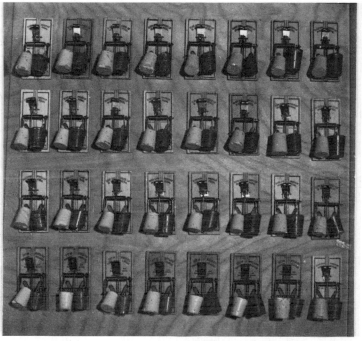

Fig. 4-2 Model of a critical mass of fissile nuclei.

Suppose we now make a very small array, one consisting of only a few traps but with the same spacing between them as before. In this case the two original corks that fly up will be apt to fly out of the array and miss the other traps, so there will be no reaction. Think of the traps as fairly densely spaced, so that a very large array at this spacing would experience an explosive reaction. There must be some intermediate size in which so many corks are lost popping over the edge of the array that, on an average, the two corks from one trap spring only one other trap. This size of array is just on the verge of being large enough to sustain an explosive reaction. It is a "critical size" of the array of mouse traps. Analogously there is a "critical mass" of a sphere of uranium at its normal density. A sphere larger than that will sustain an increasing reaction, one smaller will not.

If the array is very large and if the area of the trigger is large enough compared to the area per trap, so that two corks on the average spring more than one other trap before they stop bouncing around, then the rate at which traps are sprung will go on increasing until many of the traps have been sprung. This is a model of an exponentially increasing chain reaction. The rate at which corks are thrown into the air is about proportional to the number of corks already in the air.

Suppose instead we keep the original number of traps but increase the scale of the checkerboard, spreading the traps farther apart without changing their size or the size of their triggers. Then there is more space in which the corks can land without triggering a reaction. For sufficiently low density, the two corks from one trap will spring less than one trap on the average, and the intensity of the reaction will decrease (again exponentially). Thus the chain reaction may either grow or die out depending on the density with which the traps are packed together. The denser the material and the larger the fission cross section, the easier it is for the reaction to grow and the smaller is the critical mass.

In the mouse-trap model gravity turns the flying corks around so they come down and hit triggers in the same layer from which they started. Inside uranium metal the neutrons fly straight ahead through many layers of uranium atoms. The fission cross section of the tiny nuclei at the centers of these atoms is small compared with the area between the nuclei. The crucial question in this case (analogous to whether a cork will hit a trigger) is whether a neutron will hit and cause fission in a nucleus as it passes through many layers and before it gets to the outside of the uranium sphere, or before it meets some impurity atom that might absorb it without causing fission. The analogy with the mouse traps still holds: if the uranium nuclei are squeezed closer together, the neutron is more apt to hit one as it passes a given layer, and the bigger the uranium sphere the smaller is the percentage of neutrons that are lost to the outside. The criticality of a sphere of uranium thus depends on its size and its density in much the same way as for the array of mouse traps. There are correspondingly two ways to make an atom bomb explode with a rapidly increasing chain reaction, either by suddenly increasing the amount of material assembled or by suddenly compressing the material, as will be discussed further in Chapter 8.

SLOW-NEUTRON CHAIN REACTION

Most of the nuclear reactors that have been built make use of slow neutrons to produce fission because the nucleus ^{235}U has a large probability of intercepting a slow neutron as it goes by and then fissioning. In the jargon used, ^{235}U has an exceptionally large "cross section" for slow neutrons. This means that if we were to project slow neutrons at such a nucleus and count the hits and misses, we would find that it seems to be a large and easily hit target with a large cross-sectional area. That is, large compared to ^{238}U. The ^{235}U cross section is actually almost 10^{-21} cm^2 (larger than the actual area of the target nucleus because of the intrinsic uncertainty of position of a neutron with very small momentum). But ^{238}U is a different case; it can be made to fission only by fast neutrons—its slow-neutron fission cross section is zero. The slow-neutron cross section of ^{235}U is about 2000 times the cross section of ^{238}U for fast neutrons. Fast neutrons, however, have the advantage of being less absorbed than slow neutrons by other materials.

Most of the reactors in operation today are slow-neutron reactors. The neutrons as they come directly from the fission process are fast neutrons and must be slowed down. This is done by letting them collide with light nuclei. An analogy here is helpful. If a light rubber ball hits a heavy billiard ball head on, the billiard ball recoils slightly and takes a little of the KE, but the light ball bounces backward with most of its original KE. If, however, a billiard ball (not rotating) hits squarely another billiard ball, the first ball stops and sends the other one on with all the KE. This is discussed in Appendix 8, and the result is given by the formula relating the KE of mass m before and after the collision:

$$KE_{after} = \left| \frac{M - m}{M + m} \right|^2 KE_{before}.$$

This means that if a neutron of mass m collides head on with a nucleus of mass M, it loses very little of its kinetic energy if M is much greater than m and loses a very large portion of its kinetic energy if M is nearly equal to m. The same general distinction holds if the neutron bounces off at an angle, though it never loses all its KE in that case. Thus it is advantageous to arrange that the neutron bounces several times off quite light nuclei, being slowed down at each step, before encountering a uranium nucleus. This is what happens in the slow-neutron chain reaction shown in Fig. 4–3, which takes advantage of the large fission cross section for slow neutrons. It would seem to be desirable to choose protons, ordinary hydrogen nuclei, as the lightest

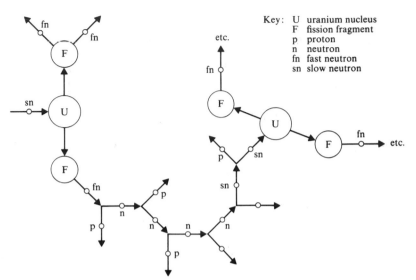

Fig. 4–3 Slow-neutron chain reaction.

available nuclei for the neutrons to hit. These are used as the "moderator" to slow down the neutrons in some reactors, in the form of ordinary water, H_2O.

However, protons are not the ideal choice, because a small percentage of the time when a neutron hits a proton it is captured to form a deuteron in the first of the reactions shown in Table 3–3. This means that the neutron would be lost for purposes of causing fission. Since about 20 collisions (bouncing off at an angle) are required to slow the neutrons down from something like 1 Mev to less than 1 ev, this can be a serious problem. Deuterium is better than ordinary hydrogen because it is less apt to absorb the neutrons, in spite of the fact that more collisions are required to slow down the neutrons. In a few reactors "heavy water," D_2O, containing deuterium is used as the moderator rather than the more abundant "light water," H_2O. A liquid like water serves a double purpose: it also serves as a coolant to carry away the heat produced in the reaction.

On the average about 2.6 neutrons are produced by each fission process. In order to sustain a chain reaction it is necessary that on the average at least one of these neutrons be captured by another uranium nucleus and cause fission. If the average number that do this is just a little more than 1, the number of neutrons bouncing about in the reactor and the rate of production of fissions will gradually increase. The increase is of the same nature as the growth of power consumption in a country or the uninhibited growth of a population—the more there is already, the faster is the rate of growth. Thus the growth of intensity of the chain reactions follows an exponential curve similar to that shown in Fig. 0–2. If the average number of neutrons from a fission causing the next fission is just less than 1, the curve slopes downward rather than upward and the chain reaction gradually dies out. In a volume filled with a mixture of uranium and a moderator to slow the neutrons down, some of the neutrons will escape from the sides of the volume. As more of the mixture is added and the volume increases in size, relatively fewer of the neutrons escape. For a small volume the fraction of neutrons causing fission is less than 1. As the volume of a suitable mixture is increased, the "fission ratio" becomes exactly 1. This amount of material is known as the critical mass as we saw in the mouse-trap analogy. For larger amounts the chain reaction will proceed; for smaller volumes it will not.

In order to achieve a critical mass at all, it is necessary to have a suitable mixture. The mere 0.7 percent of ^{235}U in natural uranium means that it has about 140 times as much ^{238}U as ^{235}U. Nuclei of ^{238}U occasionally absorb a neutron hitting them. It is therefore favorable to have enriched uranium with more than 0.7 percent of ^{235}U and thus a lower proportion of ^{238}U so as not to lose too many neutrons this way. Higher enrichment also means higher density of ^{235}U, so fewer neutrons stray out without producing fission. Commercial reactors using ordinary water as moderator and coolant commonly use uranium enriched in ^{235}U. With a smaller ratio of useless ^{238}U to useful ^{235}U, fewer neutrons are lost by absorption in ^{238}U so that one can afford to lose some in the moderator. A few reactors use natural uranium with heavy water, so as to reduce the absorption of neutrons in the moderator instead, thus avoiding the expense of the enriched uranium. As we have learned, it is easier and cheaper to separate the hydrogen isotopes then the uranium isotopes. Large reactors of this type are used to produce plutonium at Savannah River, South Carolina.

But there is another substance that does a still better job of slowing down neutrons. This is carbon, $^{12}_{6}C_{6}$, which absorbs practically no neutrons at all. In spite of the need for many more collisions because less energy is lost per collision, it does a more economical job of slowing down the neutrons. This was used in the historic reactor in which the first chain reaction was produced, to signal the start of the atomic age, in December of 1942 by Enrico Fermi and his collaborators in Chicago. They assembled a large spherical volume about 15 feet in diameter containing intermingled blocks of pure graphite, or carbon, and bars of natural uranium. At that time, no separated isotopes were available in quantity.

That first experimental reactor was a solid block with no provision for carrying away heat. It could accomplish its experimental purpose while being permitted to heat up a little without producing useful power. Larger reactors moderated with carbon and cooled by circulating air or other gas or by water have since been constructed, in this country at Hanford, Washington, for producing bomb materials and in England for producing power.

REACTORS AND THEIR CONTROL RODS
A water-moderated reactor of the type in commercial use in the United States consists of a strong pressure vessel containing an array

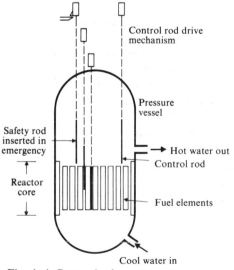

Fig. 4–4 Pressurized water reactor.

Fig. 4–5 A typical fuel assembly, consisting of many fuel rods, being lowered into place in a reactor.

of very carefully arranged vertical fuel rods of enriched uranium surrounded by water that is flowing through, as sketched in Fig. 4–4. A bundle of fuel rods being installed is shown in Fig. 4–5.

The water serves a dual purpose. It acts as moderator, slowing down each neutron as it emerges from one rod and before it enters another to cause fission (if it is not absorbed or does not escape out of the core of the reactor in the meantime). The water also acts as a coolant, carrying away the heat produced in the fuel rods by the fission. The fuel rods get very hot and would melt if they were not cooled by the water, which carries the heat away to where it can be used. The uranium is separated from the water by a very thin casing or "cladding" that lets the heat pass through but strongly suppresses the amount of radioactivity that gets out into the water.

If this were the whole apparatus, once the chain reaction is started the number of neutrons and the intensity of the reaction would tend to increase until the water could no longer carry away the heat and the reactor would destroy itself. It is essential to be able to turn the chain reaction on and off at will, and to control its level. In these slow-neutron reactors this is done mainly by use of the nuclear reaction

$$^{10}_{5}B_5 + n \rightarrow {}^{7}_{3}Li_4 + \alpha + 2.8 \text{Mev.}$$

Here B stands for boron, Li for lithium and α for the helium nucleus $^{4}_{2}He_2$, the alpha particle. This is quite similar to the fission reaction of much heavier nuclei, and the reaction products fly apart energetically in the same way. This reaction also has the important property that in it the boron absorbs neutrons with an exceptionally large cross section for slow neutrons. Both properties are used in controlling reactors. The control rod sketched in Fig. 4–4 is made of boron in order to absorb neutrons. The rate at which it absorbs neutrons, and thus prevents them from causing further fission, can be adjusted by varying the height of the control rod. In this way the reactivity of the reactor, or the intensity of the chain reaction, can be controlled. Actually a dozen or so control rods are needed, and in many reactors the same set acts as both normal control rods and safety rods, shutting down the reactor on emergency command. The control-rod mechanism of a pressurized-water reactor, with its pressure vessel partially disassembled, is shown in Fig. 4–6.

Fig. 4–6 Control-rod driving mechanism of a PWR, partially disassembled.

The other property of this reaction—the fact that the two charged particles fly energetically apart—is used to signal when the control rods should be inserted or withdrawn. Any electrically charged nucleus, such as this Li or α tearing through matter at high speed, knocks electrons out of the atoms or molecules it passes and leaves a path of ionization in its wake, positive ions and lone electrons that take some time to recombine. If this happens in a photographic film or emulsion that is then developed, the path appears as a black track, which provides one simple way to "see" these particles. One can even "see" in a negative way that the above reaction is caused by a neutron and not a charged particle for there is no incoming track.

Seeing individual particle tracks in this way is important for research purposes and for assuring us that such reactions really happen, but in a reactor the ionization from many tracks is detected electrically instead, in a simple monitoring instrument known as an ionization chamber. This instrument gives an instantaneous indication of the density of the neutrons causing the reaction. As shown in Fig. 4–7, this instrument consists of a bottle filled with a gas containing boron (usually BF_3)

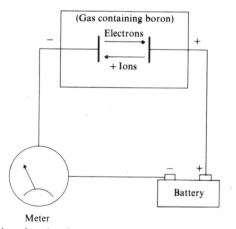

Fig. 4–7 Ionization chamber for measuring passage of slow neutrons. Boron-10 in the gas absorbs a slow neutron to make a Li-7 nucleus and an alpha particle that fly apart energetically and ionize many molecules of gas. The positive ions and free electrons so created migrate toward the electrically charged plates. The current read by the meter indicates the number of neutrons flowing by.

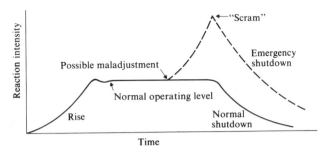

Fig. 4–8 Reactor power varying with time, normally and abnormally.

and two plates connected by electric wires to a battery and a sensitive meter to measure the electric current that flows through the chamber when the free electrons and ions created by the ionization process are pulled to the electrically charged plates of the ionization chamber (the electrons to the positive plate and the positive ions to the negative plate). An instant indication of the neutron intensity is given by the reading of a meter measuring this current, and this reading is used as a basis for moving the control rods slightly to keep the intensity of the reaction fairly steady. The current can also be used, when it gets too strong, to trigger the emergency control rods and "scram" or shut down the reactor. This is an important safety device. Such a sequence of events is suggested in Fig. 4–8.

WATER-MODERATED REACTORS

A nuclear electric power plant of course involves other apparatus in addition to the reactor. There are two important types of plant that use ordinary water to slow down the neutrons: the pressurized-water reactor and the boiling-water reactor. The pressurized-water reactor is sketched in Fig. 4–9. It makes use of the fact that the temperature at which water boils depends on the pressure, and under very high pressure very hot water can be contained as liquid, not converting into steam. The diagram shows where the water, under very high pressure, circulates at perhaps 600 degrees on the Fahrenheit scale. (The shape of the curve in Fig. 2–6 of Chapter 2 shows that it takes very high pressure indeed to make the boiling point higher than this.) This circulating water carries heat from the reactor core at the left to the heat exchanger. The water outside the coils in the heat exchanger is under only fairly high pressure and boils. This heat exchanger is then a boiler, or steam generator. The steam produced runs a turbogenerator, as shown in Fig. 4–9, just like that in a coal-fired plant with its boiler. The steam is condensed back into water in another heat exchanger that is cooled by circulating river water through it. The river water is thus heated and pumped back into the river, heating the river.

The "boiling-water reactor" differs from this by having the reactor itself act as the boiler, eliminating the need for the first heat exchanger, as in Fig. 4–10. The same steam that comes from right next to the uranium-containing tubes in the core goes through the turbine. Since it becomes radioactive from slight leaks in the thin "cladding" of the uranium (and by radioactivity induced by neutrons just outside the cladding), great care must be exercised to avoid leaks in the turbine. This will be discussed in more detail later.

The various types of power-producing and plutonium-producing reactors may be classified as follows:

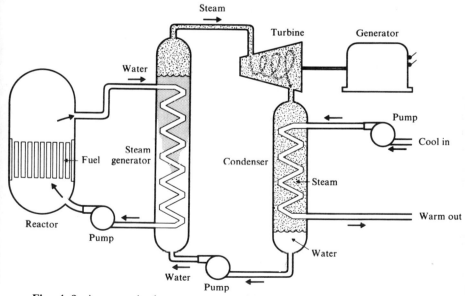

Fig. 4–9 A pressurized-water reactor with heat exchanger (left), supplying steam to the turbine with its steam condenser (right).

Thermal neutron	Natural U Carbon-moderated	(Early type, some air- or water-cooled at Hanford, Wash., and in England)
	Natural U D$_2$O-moderated	(Savannah River, some research reactors and others)
	Enriched U H$_2$O-moderated	Pressurized water: submarine, industrial, Shippingport, Pa., Rowe, Mass. Boiling water: Dresden, Ill., Vernon, Vt., and others
Fast neutron	Enriched U Liquid sodium-cooled	Breeder reactor Fermi reactor near Detroit

Altogether there were in 1971 about 20 operable nuclear power stations in the United States, and reactors then on order or under construction were expected to bring the number up to about 100 within a few years. In Great Britain the number in 1971 was almost as great and with a longer record of having produced much more power, as indicated by

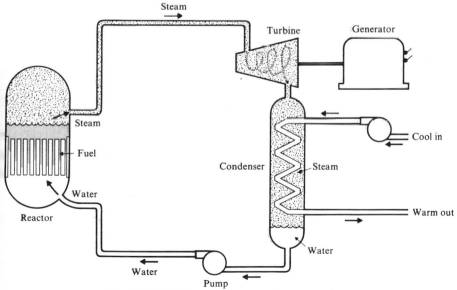

Fig. 4–10 Boiling-water reactor for comparison.

the figures in Appendix 14. There are considerably smaller numbers in each of several other countries including the Union of Soviet Socialist Republics.

Much of the earlier experience with pressurized-water reactors came from nuclear-powered submarines, where cost was not a limiting factor and for which nuclear power, with no demand on air for burning, is particularly well suited. There are 41 nuclear submarines in the United States Navy (each capable of launching 16 long-range nuclear missiles) and a considerably smaller but growing number in the Soviet Navy. Two nuclear-powered merchant ships have been built. The United States ship *Savannah* operated for a year on a demonstration basis, heavily subsidized by the government, and was then retired from

Fig. 4–11 A pressurized water reactor in practice divides its energy between two or more heat exchangers, in place of the single one in Fig. 4–9.

service as uneconomic. The West German ship *Otto Hahn* was put in service in 1968 and operates as an ore carrier.

BREEDING

The principle of "breeding" new nuclear fuel is based on the fact that there are two man-made isotopes that are somewhat similar to ^{235}U in having a large fission cross section for slow neutrons. These are ^{233}U and the plutonium isotope ^{239}Pu. The latter is made from ^{238}U and since ^{238}U is 140 times as plentiful as ^{235}U (in natural uranium), this conversion raises the possibility of vastly increased sources of fuel. The reaction by which Pu is produced is this:

$$^{238}_{92}U_{146} + n \rightarrow {}^{239}_{92}U_{147} \xrightarrow[20 \text{ min}]{e^-} {}^{239}_{93}Np_{146} \xrightarrow[2 \text{ days}]{e^-} {}^{239}_{94}Pu_{145}$$

(Here Np stands for neptunium.) A fast neutron is absorbed and then two successive electrons (labeled e^-) are emitted in two successive processes of beta decay. The first beta decay takes roughly 20 minutes on the average and the second two days. The neutron adds one unit of mass, so that A increases from 238 to 239, and the two departing electrons add two units of charge, so that Z rises from 92 to 94. The cross section for this reaction is much greater for fast neutrons than for slow neutrons, but still only about one percent of the cross section of ^{235}U for slow neutrons.   Hence this reaction takes place in a fast-neutron reactor with no moderating material. A similar reaction produces ^{233}Th from another metal, thorium, that occurs rather extensively in India and, in less concentrated form, in the granite of the White Mountains of New Hampshire, for example.

Plutonium has neutron properties rather similar to ^{235}U, but its metallurgical properties make for greater difficulty. Getting the heat out from the inside of the metal pellets to the cooling water is a problem, for the heat conductivity of plutonium metal is very low. The other problem is that plutonium is even more susceptible than most metals to radiation damage, when fast charged particles, such as the fission products, fly through the metal and its atoms, knocking things out of their way as they go, not only disturbing electrons but also dislocating nuclei. Such radiation damage arising from the presence of neutrons causes metals to crack and swell in troublesome ways and makes it difficult to use plutonium as a reactor fuel. Experiments are being conducted on the use of plutonium oxide, rather than plutonium metal, as a small fraction of the fuel in some reactors, but up until now plutonium has been used practically only for bombs. Further research on this problem seems imperative so that the breeding process, when sufficiently developed, may significantly increase the useful supply of nuclear fuel.

SOME TYPICAL REACTORS

As illustrations of the economic and technical features of present-day reactors, owned and operated primarily by electric utility companies,

Fig. 4–12 The Yankee nuclear power station at Rowe, Massachusetts.

we may contrast two nuclear electric power plants not far from one another in central New England. The older and simpler of the two is the "Yankee" power station at Rowe, Massachusetts, in a remote location on a relatively small river, the Deerfield River, about half way between Greenfield and North Adams. (Figure 4–12.) This was built about 1960 at a total cost of $44 million. Of this, $5 million came in the form of a federal government subsidy toward the development costs. For the first five years there was a further subsidy in the form of waiving of interest charges on the valuable nuclear fuel in the reactor core, which is obtained from the Atomic Energy Commission. This reactor is of the early pressurized-water type, patterned after submarine reactors. It has the longest and most consistent record of producing useful electricity of all reactors in the United States, although two others started before it, the first near Pittsburgh in 1957. Its electrical power output is 185 Mw, megawatts, or 1 million watts or a thousand kilowatts. The power required to keep a fairly bright light bulb burning is 100 watts. One Mw is enough to keep ten thousand such light bulbs burning. The reactor at Rowe is thus capable of keeping about 2 million light bulbs burning.

One should remember that the watt and kilowatt are measures of *power* and that energy is power multiplied by time. One kilowatt expended for one hour requires one kilowatt hour, or one kwh, of energy. One kwh is consumed by 10 hundred-watt light bulbs burning for one hour or by 5 such bulbs burning for two hours. The cost of electricity is measured in cents or in mils per kwh. The Rowe reactor has been producing electricity at a cost at the generating plant of about 10 mils, one cent, per kwh. This is a fairly good performance but not as good as can be done. Particularly by using larger plants, as well as more modern design, it is possible to bring the price of electricity down. This is the economic reason for the strong tendency to go to larger and larger generating plants, both nuclear and coal-fired plants. The most economical nuclear power that has been produced to date is that from the "Connecticut Yankee" plant below Hartford on the Connecticut River. This large pressurized-water plant has produced electricity at 6.5 mils per kwh. By way of comparison, the lowest cost for a coal-fired plant has been achieved at Bow, New Hampshire, 5 mils per kwh (except that plants located at coal mines can beat this price). Thus nuclear generation has almost but not quite caught up with coal-fired generation of electricity in economic cost, if we overlook certain subsidies of the nuclear enterprise.

By way of contrast we may describe the larger reactor now being built on the Connecticut River at Vernon, in the southeast corner of Vermont, about seven miles down the valley from a small city, Brattleboro. As at Rowe, it is built beside the mill pond of an existing hydroelectric plant and was designed to draw cooling water from the pond at the rate of 800 cu ft/sec. This is about two-thirds of the annual average flow of the Connecticut River and would exceed the flow in dry periods. It has been estimated by a biologist that the heating of such water as it flows through the plant would kill about 95 percent of the microorganisms and other aquatic life in the water.

To avoid the difficulties of insufficient flow and overheating of the water, cooling towers are being provided which instead dissipate the heat into the air. In the process they evaporate water into the air as water vapor or steam, in the amount of one percent of the river flow. Since this is known to be a region of rather frequent atmospheric inversions and occasional fogs, the evaporation will probably affect the local climate adversely. The newer Vernon reactor will be about three times as powerful as the one at Rowe, being rated at 515 Mw of electric power. It is to be of the boiling-water type, but of a somewhat more advanced design than the first generation of boiling-water reactors such as one at Dresden, Illinois. A plant which was to serve as the prototype of this more advanced design was built at Oyster Creek, New Jersey, near New York City, beginning in about 1965 and was scheduled to go into operation in 1967. The plan was that the construction was to be started on the Vernon reactor and several others like it after the initial operating experience had been attained at Oyster Creek, so as to leave the possibility of proceeding with more confidence and bene-fiting by any improvements of design that might be suggested. However, the financing of the Vernon plant was of course planned in advance and once such financing is arranged there is a strong economic incentive to go ahead because of the fixed charges on the money involved.

Unfortunately the Oyster Creek plant did not perform on schedule, and it was not until about three years later that it was able to produce appreciable amounts of power. Then in its first year of presumably normal operation it was "on line" 57 percent of the time. It was thus deemed necessary to go ahead with the construction of the Vernon plant and several others without the benefit of experience with the Oyster Creek prototype. There is no telling in advance whether they will be plagued with trouble to the extent that the Oyster Creek plant has been.

There are some signs of improvement in the initial operation of reactors. Four other large reactors that went on line about the same time as Oyster Creek, in about 1970, did better, achieving about 80 percent operation in the first year. It is a young industry, and experience is being gained. But the fact that the Vernon plant was built before its prototype could operate is one indication that the industry is perhaps expanding more rapidly than is prudent. Until 1971 all of the commercial reactors operating in the United States were licensed by the Atomic Energy Commission as "experimental" installations.

The core of the Vernon reactor consists of 68 tons of uranium enriched to contain 2.7 percent of ^{235}U. This is in the form of ceramic pellets packed into 23,000 thin-walled tubes of stainless steel or zirconium each $\frac{1}{2}$ inch in diameter and 12 feet high.

Following experience with other reactors, the fuel tubes will be used until about 6 percent of the ^{235}U in them has been consumed—6 percent "burn-up." To go further would accumulate too much radiation damage and radioactivity and neutron-absorbing material. About one-third of the fuel elements will have to be replaced at the end of each year of operation. (They are not all replaced at once because some are in regions of higher reaction intensity than others in the core, and it is economical to get maximum use from them.)

The cost of the enriched uranium in the core as sold to the power industry by the Atomic Energy Commission (AEC) is about $20 million. This is probably considerably less than the actual cost of producing it and thus represents an important government subsidy to the nuclear power industry. There is, however, some question about this because the bookkeeping is not readily available and is confused by the question of how much of the cost, particularly development and plant cost, should be written off to production of material for nuclear weapons.

Uranium enriched to almost 3 percent of ^{235}U as used at Vernon and indeed in all boiling-water reactors, is the most common type of reactor fuel in the United States. That used in the fast-breeder reactor near Detroit is considerably more highly enriched, as would be the case for two other breeders in New York State. But in a few pressurized-water reactors an interesting mixture of concentrations has been employed. Instead of using uranium uniformly enriched to about 3 percent, these reactors are mostly natural uranium fuel rods, but a few of the fuel rods are made of uranium enriched to about 85 percent in ^{235}U. Enough of the highly enriched rods are used so that the entire load averages about 3 percent enrichment.

The 85 percent enriched rods are dangerous in the sense that many of them placed together could make a critical assembly that would explode, but they are nicely dispersed between natural uranium rods in the reactor vessel. The economy of this method of obtaining an average enrichment may be a by-product of the routine production of highly enriched material for weapons. Unfortunately, from the point of view of avoiding weapons proliferation, it means that the fissile material owned commercially and handled with questionable care and occasionally sent astray by commercial carriers includes this uranium of weapons-grade enrichment. (See Hosmer excerpt, Appendix 16.)

FISSION PRODUCTS: DECAY OF FISSION FRAGMENTS

In Fig. 3–13 we learned that the region of stable nuclei starts out with neutron number N about equal to proton number Z for the light elements but that the general path of the stable region curves upward as N becomes considerably greater than Z for the heavy elements. This is indicated again schematically in Fig. 4–13. We have learned that if a nucleus is formed far away from the region of stability, it will have much too high an energy and may move toward the region of stability without changing its total number of nucleons A. It does this by changing neutrons into protons in the beta decay process in which an electron is emitted (or by changing protons into neutrons if it starts on the proton-rich side of the region of stability). In Fig. 4–13 a straight line is drawn from ^{236}U to the origin where $N = Z = 0$. All points on this straight line have the same ratio N/Z as ^{236}U. When ^{235}U swallows a slow neutron to become ^{236}U and immediately undergoes fission, each of the fragments has about this same ratio N/Z and lies practically on this line. One of the fragments is always considerably heavier than the other so they may lie at the two points indicated by the crosses on the line. These two fragments might for example be the strontium and xenon isotopes indicated in the following sequence:

$$^{235}U + n \rightarrow \; ^{236}U \rightarrow \; ^{98}_{38}Sr + \; ^{138}_{54}Xe$$

$$\downarrow n \qquad \downarrow n$$

$$^{97}_{38}Sr \qquad ^{137}_{54}Xe$$

$$\downarrow \beta \qquad \downarrow \beta \quad etc.$$

The uranium nucleus does not always divide in the same way so the fragments may instead be at another pair of the small circles indicated on the line, one in the heavy group and one in the light group.

The region of the heavy group, surrounded by a circle, is shown magnified in the large circle in Fig. 4–13. Each of the fragments very quickly (in less than a millionth of a second) emits one or two neutrons on the average. The emission of one neutron is indicated by the arrow pointing downward from the lowest fragment of the three for which fission products are shown in the enlarged figure. This is normally followed by a succession of beta emissions, indicated by the succession of arrows downward toward the right along the constant-A line. Most fission fragments give rise to a succession of three or four such beta decays, each one emitting an electron indicated by the e⁻. In each of these a neutron is changed into a proton so the ratio N/Z is becoming closer and closer to the stable value and these successive nuclei are becoming more and more nearly stable. The first nucleus of this chain is very unstable and the first beta decay proceeds rather quickly, perhaps in about ten seconds. The next may take longer, several minutes or hours.

The last decay in the chain may take many days or even many years. This next-to-last nucleus of the chain, the one just before the stable end product, has the longest average lifetime and is the one that causes trouble by remaining for a long time in the fuel. It is radioactive, since it is still capable of emitting a beta ray, so must be handled with care. It is one of the radioactive fission products which must be very carefully disposed of in such a way as not to cause biological harm. Each of the chains of beta decay from one of the original fission fragments leads to such a radioactive fission product. There is quite a collection of them, some of them decaying much more rapidly than others and some more troublesome than others. Some of the most important of the fission products and their decay times or "half lives" are given in Table 4–1.

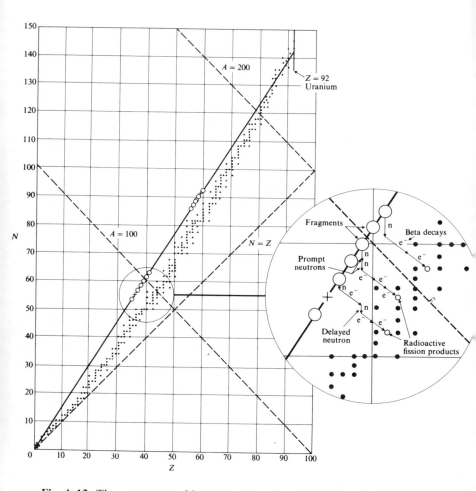

Fig. 4–13 The two groups of fragments and their decay, resulting from uranium fission. The insert shows the decay scheme of the fragments of the light group as an example. The fragments of the heavy group decay in a similar way. The decay represented by a short arrow downward toward the right means one less neutron and one more proton, as occurs with electron emission in beta decay. An arrow straight down leads to a nucleus with one less neutron, as occurs when a neutron is emitted. Prompt neutrons come directly from the fission fragments immediately after fission. Delayed neutrons are emitted following beta decay, which is an intrinsically slower process. The dots represent the nuclei that exist in nature, as in Fig. 3–13.

TABLE 4–1

Half lives of some important fission products

| Light group | | Heavy group | |
Atoms	Half life	Atoms	Half life
$^{90}_{38}\text{Sr}$	28 years	$^{133}_{54}\text{Xe}$	5 days
$^{91}_{39}\text{Y}$	57 days	$^{137}_{55}\text{Cs}$	30 years
$^{95}_{40}\text{Zr}$	65 days	$^{140}_{56}\text{Ba}$	13 days
$^{99}_{43}\text{Tc}$	10^6 years	$^{141}_{58}\text{Ce}$	29 days

Any such radioactive substance has a decay which follows an exponential curve of the sort with which we have become familiar, but sloping downward. One normally has an enormous number of nuclei of a certain kind. Any one of them may undergo beta decay at any time, completely at random. The number undergoing beta decay per second is proportional to the number of nuclei. As some of them decay, the number remaining becomes smaller and smaller. The situation is rather the opposite of the fuel-consumption curve or population-growth curve that we have discussed. In this case the fewer radioactive nuclei there are left the more slowly we lose them, and the downward exponential curve indicating the intensity of the beta decay tapers off gradually toward zero. The time it takes for the intensity to get down to half of its original value is called the half life, indicated by $t_{\frac{1}{2}}$ in Fig. 4–14. In twice this time it gets down to $\frac{1}{4}$ of the original value, etc., so the half life gives an indication of how rapidly the intensity gets down to comparably small values, but it never gets quite down to zero. This is the meaning of the half lives listed in the table above.

A long half life of a substance means that its radioactivity lasts a long time but, just because it is spread out over a long period, it is relatively weak at early times compared with the same amount of a short-lived radioactive substance. A given quantity of a short-lived substance is very intense at first because it is getting rid of its radioactivity fast. The very short-lived ones are easiest to dispose of because one can keep them confined until they almost die out. The very long-lived ones are relatively weak. They are said to have a low specific activity, or

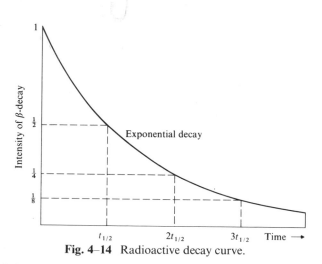

Fig. 4–14 Radioactive decay curve.

radioactivity per gram. Those of intermediate half life are the most troublesome to dispose of but prove useful in limited quantities in some medical and industrial applications.

The two elements strontium-90 and cesium-137, having half lives of about 30 years, are the most troublesome of the lot. They have short enough lives to have a very intense beta decay and yet long enough that it is difficult to wait for them to decay and become much less potent. They occur in such quantity that after 30 years, when their specific activity is reduced by one-half, or even after 100 years, when it will be down to one-tenth of its original value, they still will be so radioactive that disposal remains a serious problem. The fission products with half lives of only a few days lose their intensity rather quickly and the one with a half life of a million years, technicium-99, undergoes beta decay so gradually that the decay is very weak.

The four pairs of fission products indicated in the table are only a sample: there are about a dozen such pairs which occur with almost equal abundance. The frequency of occurrence of the various fission products is indicated in Fig. 4–15. The left-hand peak corresponds to the light group and the right-hand peak to the heavy group. The tops of the peaks are broad, indicating that there are about a dozen pairs each occurring in about 6 percent of the fissions.

REACTOR STABILITY

The decay chains from the fission fragments have another extremely

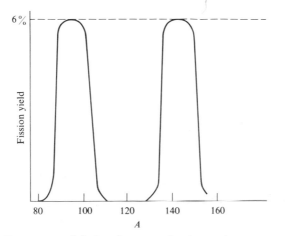

Fig. 4-15 Percentage of fission fragments having various mass numbers. For each one in the heavy group, there is a corresponding fragment in the light group.

important feature, one that is essential in questions of reactor safety. There are a few of the chains in which a neutron is emitted after the first beta decay as indicated by the vertical arrow in the middle chain of the magnified part of Fig. 4-13. As we have said, neutron emission takes place extremely quickly, while beta decay is rather slow. Thus the neutrons given off immediately by the fragments are called prompt neutrons whereas the neutron emission following beta decay is delayed by the time of the beta decay, about ten seconds on the average, and it is called a delayed neutron. In the fission of ^{235}U almost one percent of the neutrons are delayed neutrons. This is very fortunate for the problem of reactor control.

When a prompt neutron is emitted from a primary fission fragment in a slow-neutron reactor, it bounces about among the hydrogen nuclei of the moderator for about 20 collisions until it is slow enough to cause the next fission. This process takes place in a short span of about 10^{-4} seconds. If this were the whole story, the reactor would be extremely hard to control and it would be difficult to avoid a disastrous runaway of the intensity of the chain reaction. Suppose, for example, that the control rods are adjusted very carefully so that the fission ratio is 1.001. This means that from one generation of fissions to the next, just a little more than one neutron from each fission succeeds in causing the next fission, on the average, and the intensity grows by only one part in a thousand or $\frac{1}{10}$ of one percent per generation. But if a generation takes only $\frac{1}{10,000}$ second, this means that the exponential growth more than doubles in $\frac{1}{10}$ second. This would be a ticklish situation; the intensity could get out of hand before the operator or the control apparatus would have time to react.

But here the delayed neutrons come to the rescue. With them, the operator can adjust so that the fission ratio is 1.001 including the contribution of the one percent of delayed neutrons. This means that without them, with only the fast neutrons, the fission ratio would be about one percent less or only about 0.99, and the reaction would gradually die out. Thus the increase is dependent on the delayed neutrons and proceeds at their much slower pace. This gives the operator, or the automatic instruments, plenty of time to make readjustments if it is observed that the reaction is increasing too fast. But if the operator makes a mistake of more than one percent in making his adjustments, he may be in for trouble. Some of the unfortunate reactor accidents which have happened have been due to this type of trouble.

In round numbers that are easy to remember, we have said that about 1 percent of the neutrons are delayed, and delayed by an average of about 10 seconds. More specifically, the average delay time is 14 seconds. From uranium-235 about $\frac{3}{4}$ percent of the neutrons are delayed. From plutonium-239 only about $\frac{1}{4}$ percent of the neutrons are delayed. This tends to make a plutonium reactor somewhat trickier to control.

Some reactors have an additional feature contributing to their freedom from runaway accidents, depending on the fact that as the temperature gets higher the fission ratio or multiplication factor is reduced.[3,6] The best example of this is the boiling-water reactor. The neutrons have to be slowed down to take advantage of the large fission cross section of ^{235}U for slow neutrons. The slowing down is done by boiling water surrounding the fuel elements. If this water were all removed, the neutrons would not be slowed down and the reaction would stop. Around the fuel elements the boiling water consists of water containing bubbles that are rising to the upper part of the vessel that is filled with high-pressure steam. If through some accident the local temperature should rise above normal, this would rather quickly make more bubbles locally and thus expel some of the water and reduce the rate at which the neutrons are slowed down. This would reduce the local intensity of the reaction and tends to lower the temperature. This is what is called a stable or "fail safe" situation. A pressurized-water reactor does not operate in the same way because the high pressure keeps the water from boiling until the temperature is far above normal, and there could be a rather serious "excursion" of neutron intensity before boiling would occur.

REACTOR ACCIDENTS AND REACTOR SAFETY

In connection with the political and economic decision as to how fast reactor technology should be pushed and commercialized, it is important to consider the history of reactor accidents as a background for judging how likely it may be that similar or even much more severe accidents might occur in the future.[7] In discussing accidents, it must be borne in mind that great care is taken in the construction and operation of reactors to avoid accidents. The builders and operators are aware that the nuclear chain reaction has its own peculiar hazards, quite different from any others in industry, and that an extraordinary degree of caution is required. While tribute must be paid to the extent of the safety measures built into modern reactors and to the safety record which boasts that there have been no identifiable deaths as a direct result of accidents in American commercial reactors, the question remains whether the usual precautions are sufficient to prevent serious accidents in the future when many more much larger reactors pushing closer to the limits of technical feasibility will be involved. (It is thought that there may be between 600 and 1000 reactors in the United States by the end of the century.) There have been, of course, quite a number of minor accidents in the nature of personal injuries, fewer than are common in all industry, but in this case sometimes involving radioactivity. A study of these (T. J. Thompson[9]) indicates that they are typically due to either carelessness, design error, or mechanical failure. This is true also of the few more serious accidents in the worldwide experience.

The accidents which are a valuable part of the experience of past performance may help us judge whether the precautions are sufficient or whether the program would not wisely be pursued more slowly and in a way to make possible still greater degrees of precaution. In addition to the precautions of duplicate monitoring devices and elaborate shut-off procedures, the last-ditch precaution provided for most modern commercial reactors is a thick steel or concrete shell, with its conspicuous hemispherical dome, as mentioned in Chapter 2. Such an outer containment vessel is capable of withstanding enough internal pressure that a mild explosion equivalent to perhaps a ton of TNT could be contained within it, thus preventing the spread of dangerous radioactivity throughout the vicinity of the plant. Of the four accidents we shall discuss, only the one near Detroit was protected with such a shell, and the shell was not breached. Such a shell is discussed in connection with the idea of pressure on page 30.

The first of these four accidents occurred at Chalk River, Ontario, in 1952. Chalk River has a rather low-power reactor in which the neutrons are slowed down by a heavy-water moderator. Cooling tubes run through the heavy water carrying light water under lower pressure, and this absorbs some neutrons. The reactor is inherently unstable in the sense that overheating can boil away the cooling water, thus reducing the absorption of neutrons, while leaving the heavy water to slow them down, and this increases the reactivity. The accident happened when an operator in the basement turned some wrong valves—the initial act of human error. The chief operator ran down from the control room to see what was the matter and readjust the valves. He telephoned to the control desk to have buttons pressed to insert the control rods and stop the reaction. Unfortunately the wrong buttons were pressed and, by the time this error was corrected, the control rods would not work. In the course of a few seconds the sudden rise of reactivity augmented by expulsion of the cooling water melted uranium, which reacted with the remaining water to form hydrogen gas. There was an explosion which blew the core apart. Some safety devices functioned to prevent people in nearby laboratories from being seriously hurt by radiation.

Another accident took place at Windscale, a sparsely settled part of Northern England, in 1957. This reactor was even more different from modern commercial power reactors, since it consisted largely of a large cube of graphite containing uranium and was cooled by air for the production of plutonium. This accident involved the complicated manner in which radiation damage in the carbon, caused by neutrons knocking carbon atoms into unaccustomed locations and making internal strains, stores extra energy in the graphite. When the graphite becomes hot enough that these displaced atoms can slip back into place, this annealing process releases extra energy and causes further heating. A routine procedure, for heating slightly to remove this extra energy before it became dangerous, went awry and a fire of carbon and molten uranium raged inside one part of the reactor for a day or two, spewing out radioactivity, before the trouble was recognized and the fire extinguished by pouring on water. It was fortunate that the water on the molten uranium caused no explosion. Radioactivity from the accident made it necessary to dump milk supplies from grazing lands some miles around the reactor, and objectionable levels of radio-activity were encountered on the continental side of the North Sea.

A third nuclear reactor accident was much more serious in its future implications, because it happened in the first commercial fast-breeder reactor, intended as a prototype for a whole generation of breeder reactors on which the hopes for long-term nuclear power are pinned. The technique of breeding plutonium in fast-neutron reactors had been pioneered in two earlier experimental power plants, the first at Argonne National Laboratory in Illinois and the second developed by that laboratory at the reactor testing station in a remote location in Idaho. This type of reactor differs from other commercial reactors in having a considerably more compact core containing uranium more highly concentrated in ^{235}U and efficiently cooled by rapidly flowing molten sodium, which is a solid metal at ordinary temperatures. There is the added hazard that molten sodium is chemically very active if it comes in contact with air or water. The reactor has the degree of stability provided by delayed neutrons in the narrow range less than one percent above criticality but, being a compact reactor in which there is no delay time for slowing down the neutrons, it is more sensitive to the control adjustments. In order to make such an advanced-design reactor produce plutonium and electricity as efficiently as possible, the demands on the integrity of materials tend to be pushed to the limit.

The Fermi fast-neutron breeder reactor 18 miles down river from Detroit was designed to operate at 200 or 300 Mw (Fig. 4–16). After a few months of operation early in 1966 it was shut down. It was being started up again, attaining about one-tenth of full power, when the accident suddenly took place.[8] The meters indicated something wrong, the reactor quickly heated up, and within a minute or so monitoring devices sensing excessive radiation released the emergency controls and shut the reactor down. The radiation was almost entirely held within the outer containment shell, and there were no serious

Fig. 4–16 The Fermi liquid metal fast breeder reactor, 18 miles south of Detroit.

exterior consequences. The ominous nature of the accident was not in what actually happened but in what might occur if it happened slightly differently, for example in a larger reactor at higher power. A reactor contains enough fissile material to make many atomic bombs, but sufficiently interspersed with coolant and spread out so that it is not even close to an explosive condition. In this reactor the enriched uranium was impregnated with plutonium that had been produced in the previous run, and in normal operation there would be even more plutonium. In the accident some of the fuel rods melted, and afterward it was feared that there might be a dangerous lump of concentrated fuel in the reactor.

In such a situation the danger is not that the reactor will explode with anything like the full power of an A-bomb. In an A-bomb the chain reaction lasts only an instant, but long enough to make a tremendous energy release because the fissile material is assembled very rapidly. The more rapid the assembly, the greater the yield, as is discussed in Chapter 8.

A reactor builder, in obtaining a construction or operating license, must describe what he considers the "maximum credible accident" and show that he is providing effective precautions against it. For the Fermi reactor the maximum credible accident was the melting of one fuel assembly (one out of about a hundred in the core). In the accident it was found later that there had been melting in three fuel assemblies, so the actual accident was technically greater than "credible."

In the license application, by way of precaution, it had been calculated that any way one could imagine for the fuel (perhaps melted) to assemble would be so slow as to yield no more than the equivalent of about 1 ton of TNT, in an extremely mild nuclear "fizzle" as compared with multikiloton bomb explosions. The containment of the Fermi reactor was calculated to be many times stronger than needed to contain such a fizzle. Nevertheless, this would be very destructive of the reactor itself, and the resulting burst of radiation probably be highly lethal close to the reactor.

While authorities were planning how to disassemble and rebuild the Fermi reactor after the accident, there was concern lest dislodging some fuel might make the melted-down portion more than a critical mass and cause such a nuclear fizzle. As plant officials later reported, "some hair-raising decisions had to be made." Careful probing showed that the situation was not nearly critical. Despite its happy ending, the episode in this carefully supervised operation is cause for concern about other unforeseeable accidents.

It was also found that the accident had been caused by local stoppage of coolant flow by the breaking loose of a plate that had been installed as a precaution against another type of accident. Some designers have suggested that there is a limit to how many precautions can be taken—that if there are too many, they may get in each other's way. This accident is an example. Plant construction had been approved on the basis of records in the AEC which showed the errant plate as having been welded in place. It had actually been bolted in place and the bolts had failed. This is an example of the fact that the best-intentioned quality control of vital components is not always effective.

The fourth accident to be mentioned here occurred in an experimental reactor, not a commercial one, at a remote test station in Idaho in 1961. Three rather inexperienced operators, Army personnel, apparently moved a sticky control rod in an abnormal way by hand. The rod jumped a little too far, killing the three men by a burst of radiation as the reactor exceeded prompt-neutron criticality. It was a week before shielded clean-up workers could enter the building.

It must be emphasized that these accidents and as many more that might be cited all occurred in reactors rather different from the light-water-moderated reactors now being promoted. No such accidents have occurred in these reactors, whether pressurized- or boiling-water type. With these, the performance record has been good in respect to avoiding serious accident. Up to 1971 there have been of the order of magnitude of 100 reactor-years of intermittent but fairly successful operation, something like 20 reactors for an average of less than 5 years. There is no way to estimate accurately what this experience, together with the accidents in other reactors, means in the way of probability of future accident.

Before accumulation of this experience, in 1957, an expert panel centered at Brookhaven National Laboratory on Long Island was requested by the Atomic Energy Commission (AEC) to estimate the damage from a postulated accidental explosion if one of the small reactors then contemplated (about one-sixth of the power of those now being built) should disperse as much as half of its contained radioactivity into winds blowing toward a nearby city. Their report was that, if this unlikely event were to occur, it might kill about 3000 people and cause property damage up to $7 billion. By way of contrast with this estimate, Walter Jordan of the AEC's Oak Ridge National Laboratory[12] has stated his view that, while a disastrous accident is not impossible, even with today's larger reactors it would probably not be more severe than to cause 100 deaths. He suggested that this might be likely to occur only once in 10^4 reactor-years of operation, or only once in 100 years if we have 100 reactors (or once in 10 years with the possible 1000 reactors of the future). To some that opinion may seem like a long and dubious extrapolation from less than 100 reactor-years of experience during which the Fermi accident happened in a different type of reactor. Some believe the likelihood of serious accident may be ten times greater than that, some believe it is some powers of ten less. Considered opinions can differ over a wide range in the matter of accidents.

Since there have been only about 100 reactor-years of operating experience, estimates that serious accidents will occur only once in 1000 reactor-years or longer must be based not on experience but on confidence that the safety precautions will almost always work. There are technical workers sincerely devoted to making a serious accident as unlikely as is humanly possible in the many plants planned.

The lingering doubts stem from the possibility of either mechanical failure or human failure. With modern computer-oriented control the occasions for human failure can be reduced, but the human operator is given the option sometimes to override the computer. Mechanical safety measures are redundant, so that if one component fails to "scram" a reactor in an emergency, another normally will. This eliminates the danger of single-component failure. One defective electronic relay would not make an accident, but failure of two power supplies for two different sets of redundant relays might.

While the main risk of that serious accident may be from something unforeseen, there are expected occasions when emergency shutdown or scram equipment must be very sure to function properly. When a reactor is producing full power, there must be some place for the energy to go to avoid overheating and meltdown. On the average of about once a year a power station experiences a sudden loss of load when a main circuit breaker is unexpectedly tripped (perhaps by lightning, for example). A steam valve is then automatically closed to prevent racing and ruining the turbogenerator. In a boiling-water reactor, this means that the pressure in the reactor vessel suddenly increases and the steam bubbles in the core collapse. There is thus more water to slow down the neutrons, and the reactivity of the core would be dangerously increased if the emergency release of the control rods should fail to scram the reactor. This sensitivity to pressure is the opposite of the sensitivity to temperature (at constant pressure) explained on pages 110–111, and, if the energy outlet is cut off, leads to instability rather than stability.

After a reactor of any type is scrammed and there is no more fission to produce energy, there is still the very important problem of getting rid of the "afterheat" from the intense short-lived radioactivity of the fission products. This produces power at about 6 percent of the full-power rate before the scram. For a modern 1000 Mw (electric) or 3000 Mw (thermal) reactor, this means 180 Mw of thermal power, enough to melt the core in about half a minute. In case the shut-down involves a break in the main cooling ducts, emergency cooling is provided that has to be effective within this short time.

A break in the main cooling ducts in any one reactor is considered very unlikely. (It could happen through earthquake, sabotage, or materials failure.) But its consequences could be so disastrous that precautionary emergency core-cooling is considered necessary. Until 1971 reactors were designed and put on line on the assumption that the engineering basis for emergency core-cooling design was dependable. The AEC was cautious enough to initiate tests of this assumption, and a full-scale test for the system used in pressurized-water reactors is scheduled for 1973. In 1971 preliminary results from a smaller-scale test have suddenly raised serious doubts. What was expected to be a successful demonstration turned out to be a failure. The emergency cooling water simply failed to get to the hot spots, and unexpected steam pressures ejected the emergency water from the same rupture that let the normal cooling water escape. Preliminary evaluation seems to indicate that there is insufficient knowledge of how water behaves on being forced into very hot spots.[13] This puts in doubt also the rather different emergency core-cooling systems of boiling-water reactors. Pending further evaluation, reactors continue on line in the hope that such an emergency will not arise. This can be interpreted as a further indication that the development of nuclear power plants has proceeded more hastily than is prudent.

American industrial reactors may be said to have three lines of defense between the nuclear fuel and the atmosphere. The thin cladding of the fuel elements normally confines almost all the fission products. The primary pressure vessel provides very strong containment, and it should hold even a partial meltdown caused by obstruction of a cooling duct, for example.

The third line of defense is the outer containment vessel, a large strong steel or steel-and-concrete enclosure. Even if the primary pressure vessel of a water-moderated reactor should rupture, or otherwise release its contents of hot water and steam in an emergency, the steam would build up a lower pressure in the larger outer containment, and provisions are usually made for keeping this pressure low by passing the steam through water. This also washes out much of the radioactivity carried by the steam. This would be important if some steam might have to be vented as the pressure built up during several days of carrying off the afterheat. From the safety point of view, commercial reactors in the United States are required to be more conservative than those in England and the Soviet Union, which do not have the outer containment as a last-ditch defense.

One type of accident that, it is considered, could breach the outer containment is a complete meltdown following loss of coolant and failure of the emergency cooling. In this case a ton or more of molten fuel, kept very hot by the afterheat, would drop and melt its way through the bottom of the containment and deep into the earth (this is called the "China syndrome"). Groundwater would be contaminated, and gaseous fission products would escape to the atmosphere. It is presumably because the escaping fission products are only a fraction of those produced that the number of deaths in a nearby city from such an accident can be estimated as only a hundred or so for this type of accident rather than thousands.

The liquid-metal fast-breeder reactor has its special characteristics and special problems. From the safety point of view it has two advantages as well as some disadvantages. Perhaps the greatest advantage is that the liquid-sodium coolant has a normal boiling point above the operating temperature so need not be kept under high pressure. There is no need for a primary pressure vessel, and in one design the sodium fills a very large and strong pot under an inert atmosphere beneath a thick concrete lid.

The second advantage is "Doppler broadening," an effect that stabilizes the level of the chain reaction. In normal operation, the fast neutrons must have a very special velocity (relative to that of the atoms) to be absorbed by a fissile nucleus without causing fission, and these are lost to the chain reaction. So long as the atoms are at normal temperature without much motion of thermal agitation, only neutrons of a sharply defined energy are lost this way. But if in an accident the reactor becomes hotter, the atoms in their thermal agitation have a higher velocity to be added to or subtracted from that of the neutrons. This means that more neutrons, in a wider band of energy, are absorbed, thus decreasing the reactivity and the power that was heating up the reactor. This is a very important feature and will be more important in the case of plutonium, on which liquid-metal fast breeders will eventually be run, because plutonium has only about one-third as many delayed neutrons as does uranium (about 0.23 percent of neutrons are delayed), and this provides only a narrow margin for adjustments below a prompt critical condition. Liquid sodium is a good heat conductor and transports heat well.

Among the special disadvantages of the liquid-metal fast breeder are the facts that (1) unlike water-cooled reactors, its fuel is about as highly enriched as in bombs, (2) its core is much more compact, making great demands on coolant properties and flow rate, (3) the very large amount of liquid sodium is highly combustible and would burn fiercely in air or water if its inert atmosphere should fail, and (4) bubble formation in the coolant increases reactivity. Point (4) depends on the fact that, although the neutrons need not be slowed down and remain fast neutrons, the coolant is slightly absorbing to them. If blockage of coolant flow should cause overheating and bubble formation, this would remove some of the absorbing material, increase the reactivity, and cause further sudden heating that could melt the fuel. It is problematical whether the flow of molten fuel would reduce the reactivity and leave only the afterheat to melt more fuel, or whether it would increase the rate of reaction until the core blew itself apart in a nuclear fizzle of the sort that has been discussed in connection with the Fermi reactor. In any case, the large amount of sodium and the strength of the pot containing it are considered ample to handle the heat and the pressure, and the bottom is to be designed so that molten fuel dripping on it should be spread out thin enough to be cooled.

There are other types of prospective reactors, as mentioned

in the article by G. T. Seaborg and J. C. Bloom in Appendix 15. Unfortunately, all the financial support is now going into the liquid-metal variety; unless research is reoriented we may never know whether some of the others hold superior long-range possibilities. Among them one of the most promising is the molten-salt breeder that breeds thorium into ^{233}U. From the safety point of view, one great advantage of this breeder is that the fuel would not have the high toxicity of plutonium. Plutonium as produced in reactors is considered to be the most toxic substance known because it tends to spread in extremely small particles (smaller than most dust particles), each of which is an intense alpha-emitter, so that it causes a lot of localized cell damage where it settles in the body. This is the condition that most readily gives rise to cancer. The lungs are a common site.

Plutonium nuclei are intense alpha-emitters because they have a high nuclear charge which gives a strong outward push to the alpha particles. Naturally occurring nuclei have Z numbers only as high as 92, but plutonium, 94, and some other man-made nuclei go higher than that (see Fig. 3–13). The reason that natural nuclei go no farther is that all nuclei with charge higher than 92 are intense alpha-emitters and don't last long enough to span the billions of years from the time when our elements were created until now. When ^{239}Pu is made in a reactor, some ^{240}Pu nuclei and fewer nuclei beyond are also made, all of them intense alpha-emitters. By contrast, ^{233}U and the nuclei just beyond it are in the region of very long-lived and feeble alpha-radioactivity.

With no death directly attributable to accident in the operation of U.S. power reactors, as already mentioned, nuclear industry has a safety record that might be the envy of some other large-scale industries. It is frequently pointed out that extraordinary precautions are routinely observed in the operation of reactors and it is widely advertised that nuclear production of power is safe. In judging these statements, one must be aware of the much greater need for caution in nuclear technology because the consequences of an accident, if it should occur, are potentially far greater than in most other industries.

The accidents in other types of reactors have occurred through human failure, either in operation or construction. Operating any reactor is a matter of continuous readjustment, either manual or automatic, to keep the power level high enough and not too high. Driving an automobile down a two-way highway is also a matter of continual readjustment. A driver may have passed close by a million oncoming cars successfully and feel very safe. Once in a long while a driver falls asleep at the wheel and crashes. In a reactor there are several automatic devices to prevent a crash from being serious, if they do not at some time all fail.

On the basis of the "Brookhaven Report" in 1957 (page 117), Congress enacted an extraordinary piece of legislation concerning reactor accidents, the Price-Anderson Act. With the estimate at hand that an accident might possibly cause something like $7 billion in property damage, it was found that a consortium of private insurance companies would insure nuclear industries against damage claims for only about 1 percent of that. The Price-Anderson Act requires industry to take out that much insurance privately, then provides that the government will carry insurance for $500 million. Beyond that limit there shall be no liability; that is, the public takes its own chances in the range $500 million to perhaps something like $7000 million.

This may be looked upon as a government subsidy without which nuclear industry could not afford to operate. Competing industries have no such exemption from liability. The fine print in homeowners' insurance contracts commonly excludes nuclear damage. If Congress should ever decide that the present program of nuclear energy production is not turning out as well as was hoped and should be curtailed, rescinding these provisions of the Price-Anderson Act so as to put nuclear energy on a more strictly competitive basis with other energy sources might be one possible way of cutting back. (See statement by Senator Gravel in Appendix 15.)

In the unlikely event that the outer containment should burst either from a chemical explosion or a nuclear fizzle, the resultant spread of radioactivity could be far more serious than the radioactivity from an atomic bomb—though of course there would not be the enormous blast damage of a bomb. It would be more serious both because a typical reactor after operating a year or so contains as much long-lived fission-product radioactivity as is produced by something like a thousand atomic bombs and because there would be no mushroom cloud to carry the radioactivity high into the stratosphere. (The mushroom cloud depends on enormous heat from a very powerful explosion.) Left near ground level, the radioactivity would be much more dangerous.

The large steel or concrete dome seen at most present-day power reactors is a reassuring sight in that it means that any but an unprecedentedly explosive accident, or a molten core melting its way through the bottom into the earth beneath, will be contained and will not do serious radioactive damage to the surroundings. A serious accidental meltdown not explosive enough to burst the dome immediately might in some imaginable situations build up enough pressure that part of the radioactivity would have to be purposely released to avoid bursting the dome.

One precaution, mentioned in Chapter 2, has not been taken for reasons of economy. This is to put our modern reactors deep underground, as has been done with a few reactors in Sweden and Switzerland. Rough estimates have suggested that underground construction would add less than 5 percent to the cost of a nuclear generating plant and would provide much stronger containment in the unlikely event of a very explosive accident.[13] It would also provide some protection against sabotage or warlike attack. It would not, of course, prevent the routine release of low-level radioactivity or the possibility of accident in the handling and disposal of high-level wastes.

REVIEW QUESTIONS AND PROBLEMS

1. Most United States commercial reactors are slow-neutron reactors. What property of uranium gives these types an economic advantage over fast-neutron reactors?

2. Heavy water is used as moderator in a few slow-neutron reactors. What advantage has it over light water as a moderator? What implication does this have for the type of fuel commonly used with it?

3. In the mouse-trap model, why does large size of the array, with many mouse traps, tend to make the array supercritical for a chain reaction?

4. If this large supercritical array were spread out, with the same number of traps over an increasingly large area, why would it become subcritical?

5. What happens in a boron control rod? Some control rods have to be water-cooled. Can you surmise why?

6. What is breeding? Why does it appear to have importance for the future of the economy?

7. The cost of a nuclear power plant may run
   from $400,000 to $4 million..
   from $4 million to $40 million..
   from $40 million to $400 million..
   from $400 million to $4 billion..

8. Give examples suggesting government subsidy of nuclear power plants.

9. Why are delayed neutrons important? About what fraction of neutrons from fission are delayed?

10. What safety provisions should come into play if the cooling water circulation should suddenly stop (from a broken water line, for example) in a boiling-water reactor?

11. Why would breeding of thorium produce less toxicity as a by-product than does breeding of plutonium?

Fig. 4-17 AEC headquarters building near Germantown, Maryland. This building was located in the far outskirts of Washington at a time when much attention was being given to dispersing governmental operations for the sake of reduced vulnerability to a nuclear attack of modest size.

CHAPTER 5

Effects and Uses of
Radioactive Products

THE BIOLOGICAL UNITS SUBJECT TO RADIATION DAMAGE

Animal tissue is made up of cells in a variety of sizes but large enough to be seen in some detail in a microscope. Each cell contains the information necessary for determining its role in the construction of the animal and the means of reproducing itself and passing this information on. In some ways it is like a factory. A mass-production factory such as a toy factory has input materials and patterns for stamping out the products being manufactured. The patterns may in some cases be arranged in large rows or trays for easy handling. The cell has many parallels:

Input material: Amino acids (molecular units each containing a hundred or more atoms)

Patterns: Genes (arranged in long chains contained in chromosomes)

Product: Enzymes (and other less active proteins needed for structure).

The enzymes are long chains of something like 500 amino acid molecules each. The order in which these are arranged in the chain of one enzyme is the code containing the information of how the cell must behave with other cells to make specific characteristics of the animal. The enzymes accomplish this by governing or "catalyzing" the chemistry within the cell. But the key repository of the information is the gene that assembles the enzyme. There the code resides in the order of the DNA molecules that make up the gene. Each of these molecules can pick up a specific amino acid in the assembly process. Each human cell contains 23 pairs of structures called chromosomes, each chromosome containing many genes, and together these are the genetic code determining the many characteristics of the individual.

The two chromosomes of each pair are identical (except, in the male, for one of the 23 pairs). Growth and healing depend on the fact that a cell can divide into two cells that become identical with it. In the course of the division (fission in the biological usage of the term) each chromosome reproduces itself to make a new pair so that there are again 23 pairs in each daughter cell to carry on the genetic code.

This duplication of the chromosomes does not occur in the production of the sex cells in the gonads (testes and ovaries), and thus each sperm and each egg contains only 23 chromosomes rather than 23 pairs. Two of these cells unite in the fertilization process to make a cell with 23 pairs capable of further division and combining the genetic codes from the two parents in the next generation of the animal.

Reproduction of the genes is not always perfect. Variations known as mutations occur through both chemical influences and radiation, as well as spontaneously through thermal agitation (making it advantageous to keep the temperature of the sex cells relatively low).

The two categories of biological radiation damage are somatic and genetic. Somatic means damage to the individual exposed; genetic damage is passed on to future generations.

Radiation damage consists of disruption of at least one molecule in a gene within a cell, causing a mutation. This can occur when the radiation ionizes an atom in a molecule within the gene itself, or when it ionizes some other more plentiful substance in the cell (such as water, since cells and animals are made mostly of water) which then chemically injures a gene. The mutation may be lethal, serious enough to prevent the cell from dividing to reproduce itself and thus effectively to kill the cell, or nonlethal, damaging the cell only enough that it may reproduce itself with altered characteristics. There are so many cells of a given kind that a limited amount of lethal damage does no real harm: there are enough other cells to supply the need. The damage is in a sense drowned out by the healthy cells. Lethal damage to too many cells causes radiation sickness or death of the animal. A lethally damaged sex cell has no chance to carry on the damaged code to the next generation, so is innocuous. The nonlethally damaged cells are more serious.

Somatically, the damage may mean that the cells grow abnormally and cause cancer, for example. Genetically they can cause abnormal, sometimes hideously abnormal, offspring.

Genes are classed as dominant or recessive. A gene carrying the signal "brown eyes" is dominant. Thus the brown-eye gene of one parent will dominate if that from the other parent is blue, and their child will have brown eyes but will carry both brown- and blue-eye genes. However, two parents both with recessive blue-eye genes have blue-eyed offspring. A mutation from radiation is frequently recessive and causes no harm to the first-generation child if paired with a dominant gene in reproduction. Yet the child carries the mutant gene, to be passed on through the generations until it meets a recessive mutant, perhaps caused by radiation in an ancestor from the other side. Thus each mutation is in a sense a curse on society that will eventually make trouble in some future generation. It is largely because of this cumulative nature of radiation effects in the population that the guideline figure for permitted radiation to the general public has been set at only one-third of the figure for nuclear-industry workers.

Man has evolved in an environment imposing a "genetic load" of mutations on the reproductive process. In the past the rigors of life have had ways of eliminating the unfit. Early miscarriage is a natural way of doing the same thing. New chemicals and radiation increase the genetic load, with results that are difficult to predict. This may be particularly serious in modern societies that permit more of the unfit to survive.

RADIATION STANDARDS

The amount of a dose of radiation (such as the intensity of the radiation from an x-ray tube multiplied by the time of exposure) is measured in a unit called the roentgen. This is defined in terms of the number of ions made in a certain volume of gas. The "rad" is practically the same thing—though in detail defined in terms of ionization in tissue. The "rem" is somewhat different because it takes into account that more intense local ionization causes biological damage more than diffuse ionization, for the same number of ions. (Rem stands for "Roentgen equivalent, mammal.") One rem of x-rays or gamma rays is the same as one roentgen of them. But protons and alpha particles ionize much more intensely along a short path (perhaps a thousandth of a centimeter if they have several Mev of energy). For them one roentgen is something like ten rem, the equivalent of ten roentgens of x-rays in doing damage.

Acute somatic damage increases rather suddenly at a certain amount of radiation, although some is tolerated with little effect. For example, 100 rem causes mild radiation sickness in man but death in only a small fraction of cases. If received in one dose, 250 rem causes severe radiation sickness but few deaths; however, 500 rem causes about 50 percent deaths, following the curve of Fig. 5–1.

Chronic somatic damage behaves differently. This is the effect of continuous very weak exposure to many very small doses, much too small to show on the scale of Fig. 5–1, spread out in time. One important such effect is increased incidence of cancer in a population. For extremely small exposures it is difficult to obtain significant statistics, but what indications there are, as one investigates smaller and smaller doses, make it very likely that there is no threshold for the onset of damage. This means that there is no level such that for still smaller exposures there is no likelihood of trouble. It is assumed that this is true in setting radiation standards.

The average individual in the United States was estimated in 1963 (when there were only a few relatively small industrial reactors) to receive radiation doses from various sources as shown in Table 5–1. As is there seen, the atomic energy industry of 1963 (with six relatively small reactors) caused a very small fraction of the total, but is probably ten times as important now and may be something like a hundred times

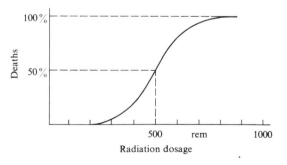

Fig. 5–1 Percentage of cases in which death occurs for various radiation doses.

as important before the end of the century if present plans are carried out, unless a recent trend toward tightening standards on effluents should become effective by then.

One sees in Table 5–1 that the average person in the United States received about 0.13 rem per year from natural sources and 0.07 rem per year from man-made sources (before 1963), making a total of about

TABLE 5–1

Radiation exposures in the United States*

	Millirems†
Natural sources	
A. External to the body	
1. From cosmic radiation	50.0
2. From the earth	47.0
3. From building materials	3.0
B. Inside the body	
1. Inhalation of air	5.0
2. Elements found naturally in human tissues	21.0
Total, natural sources	126.0
Man-made sources	
A. Medical procedures	
1. Diagnostic x-rays	50.0 (up to 95 in 1970)
2. Radiotherapy x-rays, radioisotopes	10.0
3. Internal diagnosis, therapy	1.0
Subtotal	[61.0]
B. Atomic energy industry, laboratories	0.2
C. Luminous watch dials, television tubes, radioactive industrial wastes, etc.	2.0
D. Radioactive fallout	4.0**
Subtotal	[6.2]
Total, man-made sources	67.2
Overall total	193.2

* Estimated average exposures to the gonads, based on the 1963 report of the Federal Radiation Council and adapted from I. Asimov and T. Dobzhansky, Ref. 6.
† One thousandth of a rem.
** Down to 1.0 in 1970.

0.2 rem per year. Of the man-made part, about 0.06 rem was from medical x-rays, either diagnostic or therapeutic, and only one-tenth as much from radioactivity, including both fallout from bomb tests (which reached a peak in about 1963 and is now in a slow decline) and the effluent from the few early reactors. While the table applies to exposure to the gonads, it is roughly indicative of whole-body doses.

The present level of additional radiation (over and above natural background plus medical exposure) considered acceptable for the population at large is 0.17 rem/year. This is recommended as a guideline (because of lack of authority to set a strict standard) by the Federal Radiation Council and has for years been accepted by the AEC as the maximum additional radiation to which the public at large should be subjected as a result of AEC-sponsored activities. Some activities are now held to a limit lower than this. This 0.17 rem/year is nearly equal to the natural background plus medical x-rays. Although many considerations came into play, it seems that the limit was some what arbitrarily set so that this might be so, that is, that the permitted addition should not be greater than the background radiation already being received.

The guideline limit has gradually been reduced to this figure during the years as a result of increasing knowledge of the effects of radiation. In the early days of x-rays, early in the century, radiologists allowed themselves something like 100 roentgens per year, which then seemed to be sufficiently far down on the curve of Fig. 5–1 if one worries only about radiation sickness. Experience with cancer among these workers, as well as among miners and watch-dial painters who ingested radium, led to successive reductions of the limit considered prudent.

Another group of people whose tragic exposure to radiation provides valuable information for judging the effects of radiation are the survivors of the bomb bursts at Hiroshima and Nagasaki. In the early sixties when the guideline figure was set, the data from this source seemed to indicate that moderate levels of exposure increased the incidence of leukemia (cancer of the blood) by some considerable percentage but did not affect other forms of cancer. Leukemia is a comparatively rare form of cancer so this seemed to be rather re-assuring. It has since been recognized that this was a mistake because leukemia appears sooner after exposure—has a shorter latency period—than other forms of cancer and was the only one that had yet appeared after the bombings. In the meantime the more prevalent forms of cancer have been showing up among these victims, and it is believed that the percentage increase in all forms, due to the bombing exposures, is about the same. The Gembaku hospital reports that 70 victims died in 1970 from effects of the Hiroshima bombing in 1945 (as compared with 71 in 1969).

A warning that low levels of radiation in a large population could cause very serious amounts of damage, particularly genetic damage, was issued and widely publicized at the time of concern over the atmospheric bomb tests in the late 50's by Linus Pauling, winner of Nobel prizes for both chemistry and peace. (The only other person to win two was Marie Curie, after whom a unit of radioactivity has been named.) In *No More War*[1] Pauling said:

> A single large superbomb, like the one which was detonated by the United States on 1 March 1954, causes an incidence of disease such as to lead to the death of 10,000 people by leukemia and bone cancer and possibly also 90,000 more by other diseases, a possible total of 100,000 deaths.

The full extent of Pauling's warning was not everywhere accepted. Then and even considerably later it was argued by some that there is a threshold level of radiation below which no damage is done. However, in more recent radiation standards, it was generally recognized that in setting such a limit one should realize that there is no valid evidence for a threshold and should conservatively assume that the probability of damage is linear, or proportional to the dose with a straight-line graph right down to very small doses. It is thus recognized that any radiation is bad—very small amounts contribute at least to the probability of genetic damage—but that the probable harm must be weighed against the probable benefit from carrying out the process leading to the exposure. For example, a pelvic x-ray before childbirth, or a series of them, contributes a probability of the order of $\frac{1}{10}$ percent that the child may die of leukemia, but is considered worth the chance for the sake of avoiding serious complications of delivery.

Similarly, in the days of atmospheric testing of nuclear weapons before the limited test ban of 1963, it was largely recognized that each large nuclear weapon tested in the atmosphere would through its fallout cause some number of deaths from cancer and genetic effects spread inconspicuously throughout the world. But, in the opinion of those making the decisions, the tests contributed to the probability of avoiding a nuclear war that would involve enormously many more deaths if it should occur. Thus, on balance, the tests could seem beneficial.

A prominent consultant on nuclear policy once remarked that although the numbers suggested by Linus Pauling may be high, one knows when he carries out a big nuclear test in the atmosphere that he is probably killing some hundreds of people sometime, somewhere. If he knew personally who those people were, could name them or individually identify them, he would not press the button even though he was aware of the national-security reasons for the test. Not knowing them, he does.

It was during this period and largely with this justification in mind that the guideline figure of 0.17 rem per year, about equal to natural background, was set. According to the estimates of Table 5.1, the average exposure to fallout was only about $\frac{1}{40}$ of this, 0.004 rem/year, but the guideline figure was still sometimes exceeded locally by above-average exposure of populations either near the testing grounds or at places where wind or rainfall patterns concentrated the fallout.

Various authorities take various attitudes toward the adequacy of that guideline limit. In the "Understanding the Atom" series put out by the United States Atomic Energy Commission (and available through its division of Technical Information, Box 62, Oak Ridge, Tennessee) there are two interesting booklets bearing on the subject.[6] One is called "The Genetic Effects of Radiation" by Asimov and Dobzhansky. It contains a very informative as well as authoritative discussion and concludes with the following remarks:

Some geneticists have recommended that the average total exposure of human beings in the first 30 years of life be set at 10 rads. Note that this figure is set as a maximum. A total exposure of 10 rads might increase the overall mutation rate, it is roughly estimated, by 10%. This is serious enough, but is bearable if we can convince ourselves that the alternative of abandoning radiation technology altogether will cause still greater suffering. A 10% increase in mutation rate, whatever it might mean in personal suffering and public expense, is not likely to threaten the human race with extinction, or even with serious degeneration.

The human race as a whole may be thought of as somewhat analogous to a population of dividing cells in a growing tissue. Those affected by genetic damage drop out and the slack is taken up by those not affected. If the number of those affected is increased, there would come a crucial point, or threshold, where the slack could no longer be taken up. The genetic load might increase to the point where the species as a whole would degenerate and fade toward extinction—a sort of "racial radiation sickness." We are not near this threshold now, however, and can, therefore, as a species, absorb a moderate increase in mutation rate without danger of extinction.

What humanity must do, if additional radiation damage is absolutely necessary, is to take on as little of that added damage as possible, and not pretend that any direct benefits will be involved. Any pretense of that sort may well lure us into assuming still greater damage—damage we may not be able to afford under any circumstances and for any reason.

Actually, as the situation appears right now, it is not likely that the use of radiation in modern medicine, research, and industry will overstep the maximum bounds set by scientists who have weighed the problem carefully. Only nuclear warfare is likely to do so, and apparently those governments with large capacities in this direction are thoroughly aware of the danger and (so far, at least) have guided their foreign policies accordingly.

The other AEC booklet, "Your Body and Radiation" by Norman A. Frigerio, discusses the somatic effects of radiation and their clinical symptoms and treatments, along with brief mention of radiation therapy. Without discussion of tolerable limits for the population, one viewpoint is expressed by this, "Radiation can be as helpful as harmful, and far more lives have been saved by it than lost to it." This estimate probably refers to losses by clinically identifiable cases of death caused by radiation.

A more critical opinion of the guideline figure was expressed by two investigators in testimony, dated Nov. 18, 1969, before a sub-committee on pollution of the Senate Committee on Public Works and later in two books and elsewhere.[9] J. W. Gofman and A. R. Tamplin of Berkeley and Livermore, California, made this report as a conclusion of seven years of study carried out with the initiative and support of the Atomic Energy Commission. Their conclusion, however, would require a drastic modification of AEC-sponsored activities and was not readily accepted by the commission. Their studies concern, mainly, various forms of cancer, including leukemia. They conclude that the guideline figure permitting a general-population exposure of 0.17 rad for a year is too high and should be reduced by at least a factor 10.

Their reasoning is approximately as follows: It takes about 100 rads, or somewhat less, to double the spontaneous rate of incidence of cancer. This is presumed to mean that 1 rad will increase the incidence rate by about 1 percent. Cancer induced by low levels of radiation does not show up immediately, since there is a latency period of about 20 years, so one must think of effects over the long term. For cancer production the effects are presumably cumulative, so adults more than, say, 30 years old could be affected. If the population were exposed to the guideline figure of 0.17 rad or $\frac{1}{6}$ rad per year, this means 5 rads in 30 years, or about 5 percent increase over the normal incidence. The normal incidence in adults is about 3 cases per 1000 people per year. An increase of 5 percent of that means an increase of about 15 cases per 100,000 people per year or 15,000 additional cases per year among the 100 million people in the United States over 30 years old. When the authors get down to details, they believe the actual number would be about twice this, or 30,000 additional cases of cancer per year. This they consider far too high a price to pay for "cheap" nuclear electricity.

It may be argued that the exposure these authors assumed is unrealistic. Effluents from nuclear plants are limited so as to keep the average exposure of residents near the plants well below the guideline figure. Also, not all the population would receive dosages near this level, and thus cancer incidence would not be that high even if the guideline limit were commonly reached in the atmosphere at generating-plant boundaries.

That the average exposure of the United States population would be much less than the guideline value can be simply estimated by assuming that the population is uniformly distributed over the country, with N nuclear plants, and that the radioactivity remains mixed in the atmosphere only to the height where it is already put at the plant boundary by the high stack. The spreading by winds in various directions then makes the average exposure decrease as the inverse distance from the plant. Since the United States is about 10,000 times as wide as a typical plant site, this means that the average exposure is reduced below the value at the plant boundaries by a factor of about $N/10,000$, which would mean down to one percent if we had a hundred plants. This rough estimate could be considerably increased by noting that the plants tend to be near cities, or decreased by assuming more mixing in the atmosphere and by calculating the radioactive decay of the short-lived components.

It can further be argued that most plants now operating are routinely releasing into the air only a few percent—in some cases less than 1 percent—of their permitted radioactive emission, so the pathological cost of cheap nuclear electricity will not be nearly as high as suggested. The point made by Gofman and Tamplin is not that nuclear power production will kill 30,000 people a year, but that if plants can be run at a level to keep the deaths much fewer than this, they should be required to do so as protection against future laxness (or else should not be built). One can be particularly concerned about future laxness when there may be very many more and larger plants and when the whole activity of the country may be so completely dependent on them that shut-downs for safety's sake would be strongly resisted by powerful vested interests. This could make it much harder to introduce appropriate standards then than now.

Reduction of the permitted figure of 170 millirems per year has been resisted by the electric utilities, and at least up until 1971 also by the AEC. Keeping emissions low adds to the cost of generating electricity. Low-level wastes have such large volume that it is costly to store them and the economical thing to do is to release them. While the routine releases from most plants have been kept far below the permitted level, there are often rather sudden nonroutine releases (sometimes called "burps") either when something goes slightly wrong or during the opening of the reactor vessel for routine change of fuel. These are permitted to go to ten times the permitted yearly-average level for short times. This would be particularly significant if it occurred at times of atmospheric inversions, confining the release in the manner of the Los Angeles smog.

The AEC in 1971 adopted a policy of keeping radioactive emissions "as low as practical." It announced its intention to issue licenses for operation of water-moderated power plants only at radioactive emission levels of 5 millirems per year, down to about 3 percent of the amount previously permitted, and to bring plants already operating down to that level soon. This is an important step in expressing official concern and recognition of the need for tighter regulation of nuclear hazards as nuclear materials become more commonplace. However, the announced reduction does not apply to fuel-processing plants (or to liquid-sodium-cooled breeder reactors), so as yet it improves matters only for the type of release that has encountered the most vociferous local opposition as new power plants have been proposed. More experience may be needed to determine what limits may be set as "practical" in other situations.

Gofman and Tamplin do not go into genetic effects in detail in their testimony, but they do discuss probable malformations and infant deaths. Under their assumed exposures there would probably be about 30,000 of these in addition to the incidence of cancer. Another long-time radiation worker who does emphasize these effects and who is far more critical of the accepted standards, in a more controversial way, is E. J. Sternglass of the Radiological Physics Department of the University of Pittsburgh.[10] He has studied the statistics of reported fetal and infant deaths in various parts of the world in relation to the carrying out of nuclear tests and to the little that is known about fallout patterns. The rate of infant deaths before the years of testing was decreasing at a fairly steady rate as a result of improved conditions of medicine and hygiene. The result of the testing must be seen as a modification of this rate. Sternglass attempts to separate testing results from other possible influences, such as migrations of underprivileged peoples, so the interpretation is a somewhat tricky business. His conclusion is that atomic bomb tests at remote locations have caused infant deaths on the order of magnitude of 1 percent of live births. These would add up over the years to distinctly less than 1 percent of world population, but still an enormous number.

It is widely recognized that infants, particularly in utero, are many times more susceptible to radiation than are adults. In view of the population explosion, 1 percent of infant deaths might be considered not serious, from some broad viewpoint. But there are two more serious implications to Sternglass's figure, if it is valid. First, there would probably be a comparable number of genetic deformations, though these are harder to trace. Thus exposure to fallout or similar contamination from reactors is not a satisfactory means of population control. It could succeed in limiting numbers but, aside from the grief involved in infant deaths, this would be no calamity. A more serious effect would be that it would reduce the quality of human stock through genetic damage. Second, an all-out nuclear war would probably cause fallout something like 100 times that caused by the nuclear tests of the late 1950's and early 1960's. If Sternglass is right, this would mean infant deaths at the rate of approximately 100 percent of live births, so the human race would simply not reproduce itself. At least, it would not if the war should spread to both hemispheres. (Winds are such that fallout from a bomb in one hemisphere does not readily spread to the other hemisphere.)

The most important reason to advocate arms control is to reduce the likelihood that nuclear war might occur. These considerations suggest an additional reason for arms control: to keep the number of nuclear weapons down so that even if nuclear war should occur, it would not terminate the human race.

The attitude of most critics is that Sternglass's conclusion has not been proved. There is a tendency for it to be believed or not according to the prejudices of the critic. Some of his claims, particularly those of effects near nuclear power plants, are based on much more fallible statistics than others and are most subject to doubt. It is imperative that this important subject be further and independently studied in an unbiased approach. In the meantime, it can reasonably be said that his general conclusion cannot be disproved and that he has performed a valuable service in pointing out the limits of our firm knowledge of these matters. If a war only *might* end the human race, we had better not have it. If infant mortality and malformations as a result of permitted exposures from nuclear power plants *might* be 10 or 100 times those surmised by Gofman and Tamplin, as Sternglass further concludes, this casts further doubt on the advisability of such a program.

The difficulty with making firm conclusions about these matters is that such low levels of radiation make a very small likelihood of trouble for any one individual, and such large numbers of individuals are needed for meaningful statistics that one cannot make controlled experiments. Until fairly recently there were considered to be no data with adequate control comparisons with unradiated populations at levels near or below the guideline figure. Some have considered it generously conservative in setting standards to use a linear relation rather than a threshold below which no damage is thought to be done because none is observed. Fairly recently the studies of Alice Stewart in London have both shown the unusual sensitivity of infants to radiation and have traced the linear relationship between exposure and damage down very close to the low levels permitted for the population.[11] Her studies are based on the statistics of leukemia and other defects in infants exposed to x-rays in utero as compared with the control population of those who were not so exposed. Dependable diagnoses of such large numbers of cases were available that the one-in-a-thousand chance that x-ray observation of the mother causes leukemia in the infant could be definitely established.

RADIOACTIVE WASTES

Since animals are sensitive to radiation and some of the most important fission products have decay times of many years, the important question arises: "What happens to the fission products produced in a power reactor?" A relatively small part is routinely released from the reactor into the environment; most of the rest is shipped off to a fuel-processing plant where the more serious problems of disposal are encountered. In terms of routine planned releases, the boiling-water reactor is a more serious offender than others. This is partly because the same water or steam that passes through the reactor core also passes through the turbine where at least slight leaks are inevitable. If the thin metal cladding which encases the uranium pellets were completely effective, the radioactive fission products would be contained in the fuel rods. As it is, there are two ways fission products can get out past the cladding. For the sake of promoting heat conduction, the cladding is made so thin (0.02 inch) that there is a little normal diffusion of fission products through it. A more serious problem arises from the fact that the cladding sometimes fails by developing leaks, but there are so many tubes of fuel in a core that it would not be practical to shut down the plant every time one fails. Thus 1 percent of faulty fuel rods are permitted before shut-down, and most of the time the water and steam come in contact with some of the uranium, receiving the radioactive fission products directly. In addition to the fission products that escape from the fuel pellets, radioactivity is also introduced into the coolant water in the core by the action of neutrons activating atoms from the core materials and impurities in the water, for there is of course a high intensity of neutrons throughout the reactor core. Zinc-65 from the zinc in the fuel cladding is a bad actor of this sort. To prevent the closed circuit of coolant water from becoming too loaded with impurities, these are removed by a water purification system that mixes them into an exterior solution to be treated as a low-level radioactive waste. This is permitted to be flushed down the river if tolerances are not exceeded, the economical way to get rid of it.

The pressurized-water reactor also produces the radioactive waste tritium in a special way. Neutron-absorbing impurities accumulate and the amount of ^{235}U fuel decreases slightly during the year or more that one loading of fuel is left in the reactor. To prevent the core from being less reactive at the end of the cycle, it is necessary to provide it with extra boron. This is done by dissolving boron in the cooling water and gradually removing it chemically during the fuel cycle. The control rods are inadequate for this purpose.

The difficulty here is that boron absorbs slow neutrons in a reaction that produces tritium, which gives off beta rays with a 12-year decay time. Since tritium is a form of hydrogen, it is either incorporated in water molecules in the liquid effluent or escapes as a gas routinely released up the stack, to be diluted in the atmosphere, along with the fission product ^{85}Kr. Krypton is also a gas that does not dissolve in water and has a 9-year decay time. A boiling-water reactor does not employ boron in the cooling water both because it has more stringent requirements on water purity to avoid encrustation and because it has less need for boron. In it, control of bubble size through pressure can in part compensate for changes in reactivity.

The license for some recent plants sets the permitted release of radioactive gas at an average of $\frac{1}{20}$ curie per second, or a total of 1.5 million curies per year. That is more than would be emitted from the world's supply of radium, so it seems like a lot of radiation. (A curie is the amount of radiation from a gram of radium and the world supply of radium is about a thousand kilograms.) To the question of how to dispose of wastes, part of the answer is that "dilution is the solution of pollution." That is, there is a great deal of atmosphere in which to dilute this radiation, most of which goes up the stack as gas. Some electric generating plants have used only a small part of their quota, while others have had to slow down operations during part of the year because of having exceeded the average earlier. The quota for some of the newest boiling-water reactors, such as the one at Vernon, Vermont, has been 20 times that much, or 1 curie per second, but such large quotas are being revised downward.

DISPOSAL OF WASTES FROM SPENT FUEL

By far the greater part of the fission products—almost all—go out of the reactor in the spent fuel. Once a year or so a reactor must be shut down for withdrawal of spent fuel. This is first stored underwater at the generating plant for some months to permit the short-lived components of the radioactivity to die out. It is then placed in special heavy caskets designed both to absorb radiation and to minimize the chances of breakage in case of a derailment in transit. It is then transported by conventional carriers to the fuel-processing plant. One of the dangers as the number of generating plants becomes larger is that a derailment or truck accident in a populated area will cause very serious contamination of the immediate locality (and of water supplies). At the reprocessing plant the fuel, which is solid, is dissolved in concentrated acid solutions. The useful materials are removed chemically. These include uranium, for only about 6 percent of the ^{235}U has been burned in each fuel cycle, and the relatively small amount of plutonium that has been produced (which may be about a gram for each megawatt-day of operation).

The most serious problem is the disposal of the radioactive wastes in these concentrated solutions.[14] Some of them are sorted out chemically according to their lifetimes, so that the short-lived products can be set aside and allowed to decay. The highly concentrated or "high-level" portion having long decay periods, including prominently ^{90}Sr and ^{137}Cs, are the most troublesome to dispose of. They are extremely hot in the radioactive sense, but they are also hot thermally, since the radioactive decay generates enough heat to cause them to boil irregularly and thus erupt into the atmosphere if they are not artificially stirred or cooled. Their 30-year decay time or half life does not mean that they are harmless after 30 years. It merely means that they are half as radioactive after 30 years or down by about a factor of 10 in 100 years, which still leaves them far too hot to handle or to dispose of by dilution in the environment. They are therefore stored in a rather intricate program of "perpetual care." There are almost 10^8 gallons of this extremely radioactive stuff in storage in huge underground tanks with arrangements for cooling or stirring them that require constant vigilance and dependable power supply. These are stored in about 200 enormous underground tanks (Fig. 5–2), the largest over a million gallons, at the Hanford Works on the Columbia River in the State of Washington and the Savannah River in South Carolina. The radioactivity is something like several hundred curies per gallon in the bulk of these wastes that are left from production of bomb-grade plutonium. In such production the fuel is left in the reactor a relatively short time before reprocessing, to avoid accumulation of ^{240}Pu (from ^{239}Pu plus one neutron) which is a fast-neutron absorbing "poison" reducing the efficiency of a bomb. The liquid wastes from future large reactors

Fig. 5–2 A "tank farm" under construction. Large tanks to hold high-level liquid wastes.

operating economically will be much "hotter" than that, about 3000 curies per gallon, because a much longer fuel cycle can be used.

Experience has shown that these tanks last about 20 years before corrosion from chemical action and radiation damage causes them to buckle and leak. Some have failed before that. There have been about 9 failures, one of which went undiscovered long enough for 60,000 gallons of high-level waste to escape into the soil and river.

Some underground aquifers carry water slowly, and it is quite possible that important underground sources of water supply could be irretrievably poisoned with radioactivity long before the trouble is detected.

One plan is to build new tanks into which to transfer the waste every 20 years "in perpetuity." It has been estimated that the cost of all this, spread over perhaps two centuries, will amount to only about 1 percent of the cost of producing the power of which this is a by-product. From the long-range point of view, there is a serious political question whether it is wise to permit the integrity of the environment to be so completely dependent on the continuity of political institutions and initiatives. There are those who consider it utter folly.

Indeed those who are responsible for carrying out this procedure are quite aware that it is unsatisfactory for the long term and are supporting research intended to find better methods of high-level waste disposal. One hopeful method is conversion into blocks of glassy solid that can be stored underground in dry regions such as salt mines and would be very little dissolved even if they should be reached by ground water. While the advantage of storage in salt mines was first suggested in 1954 and since promoted by Oak Ridge National Laboratory and the glassy-solid method has been announced as economic by the Canadian Chalk River Laboratory in 1958, this type of waste disposal by 1971 has not yet been put to practical use. Finally, however, the AEC is searching for and trying to acquire a suitable abandoned salt mine with the intention of making it a national waste repository. Such plans, long overdue, may be an example of a constructive response to recent outside criticisms.[17]

It is proposed to package the solidified (but not necessarily glassy) very hot wastes in narrow stainless steel canisters ten feet high and space them on the floor of a room in the mine far enough apart that the salt will heat up slowly (Fig. 5 3). The room can then be packed with loose salt and later the salt almost melts and seals in the canisters. The metal will soon corrode away, leaving the radioactive solids sealed in the salt itself, presumably forever. The strong appeal of salt beds is that these geologic formations have been there many millions of years without being washed away by groundwater.

Unfortunately the serious trouble has been encountered that most salt beds have already been disturbed by mining and exploratory drilling in a way to give them new access to water. The site first seriously considered and almost acquired at Lyons, Kansas, had to be abandoned because it was belatedly found to be within a few hundred yards of large cavities left filled with water by previous salt-mining operations. Since it appears inevitable that industry will go on producing radioactive wastes regardless of ability to dispose of them, it is very important that a suitably unviolated site for this otherwise promising disposal method be found.

Intermediate-level and low-level wastes are disposed of in other

Fig. 5–3 An experimental installation in a Kansas salt mine, to test the response of the salt to heating by electrically heated cylinders simulating those in which it is hoped to store high-level wastes.

ways. The intermediate-level wastes are radioactive enough to be very dangerous but do not produce enough heat to be troublesome. For example, they can be mixed into cement and injected into underground cavities.

All this chemical procedure produces large volumes of low-level waste in which to wash out the fractions not concentrated. Because of its large volume it would be costly to store it, so the economical procedure is to dump it into water courses. The AEC sets up standards for the radiation level of these effluents, intended to keep the exposures within the guideline limit. As in the case of low-level wastes from power-generating plants, the authorities operating the plant are responsible for policing themselves on this matter. In addition to the AEC-operated fuel-reprocessing plants, there is a commercially operated one at West Valley, near Buffalo, New York. Independent scientists from the University of Rochester undertook on their own initiative to check on the effluent from this plant and found it to be about 300 times the permissible limit of about 3×10^{-10} curies per liter (a liter is about a quart).   The United States Public Health Service is nominally interested in this sort of activity but is not equipped for it and does not perform this function.   This example points up the need for an active agency independent of the AEC to set standards of radioactivity and monitor them.

In response to this type of need, the Environmental Protection Agency was set up in 1970 and is expected increasingly to take over these functions.[19]   At the same time there has been an expression of possible redirection of the AEC role in these matters by the new chairman of the AEC, suggesting that he, at least, will seek to have the AEC become more protective of the public interest and less a promoter of nuclear power plants in the future.   (He probably refers to the present generation of water-moderated plants when he talks of deemphasizing promotion, for the AEC continues to stress development of liquid-metal fast-breeder reactors.)

A rational judgment should be made whether large numbers of power reactors should be built to produce enormous quantities of these lethal wastes before a safe method for handling them has been fully developed.   The reactors last hundreds of years and can be a curse to future generations if not permanently and completely isolated.   An important part of the social challenge of nuclear energy is to reap the benefits of nuclear energy without undue harm.

CONCENTRATION IN THE BIOLOGICAL CHAIN

Even if the standards are properly enforced, there is question whether the permitted level of radiation in rivers is sufficiently low.   It is set so as to avoid having more than 5×10^{-12} curies per liter in drinking water, apparently without any consideration of the way animals growing in the water concentrate the radioactivity.   There have been many examples of such concentration downstream from nuclear plants: plankton a thousand times as radioactive as the water in which it grows, oysters or little fish eating the plankton still more radioactive, and big fish, eating the little fish, with 20,000 times as much strontium-90 in their bones as in the water.

These are examples of the inadequacy of our knowledge of the ultimate destiny of the fission products being produced to supply us with electricity and bombs. Beyond the rivers and their estuaries is the enormous expanse of the sea, posing questions even there of biological concentrations of the effluents from the rapidly growing number of nuclear plants of the future.

MEDICAL USES OF RADIATION

Even before the advent of the chain reaction and man's use of nuclear energy on earth, a beginning was made in the use of radiation and radioactivity for medical purposes. Both x-radiation and the natural radioactive atoms were used for diagnosis and therapy, particularly radium and its daughter radon. These were usually inserted in needles where localized radiation was desired to eliminate malignant tissue.

The advent of nuclear reactors has greatly expanded the variety and quantity of radioactive materials available for these purposes.[16,22] Some of the fission products are useful for such applications, so a minute fraction of the high-level wastes from a fuel-processing plant can be diverted to these beneficial ends, but this does not appreciably detract from the huge amounts that must be disposed of in perpetual storage. A greater variety of radioactive isotopes for medical and industrial purposes is produced by neutron irradiation of various elements. The most widely used of these is cobalt-60, which is prepared by placing a fairly small cylinder of the normal metal cobalt-59 in the region of high neutron intensity near the middle of a special reactor that is operated for this purpose. Thus cobalt-60 is not a by-product of the nuclear electric industry, as fission products are. The fission products are used in such very small quantities that there is no need to run many reactors to supply them.

In a typical use of a radioactive substance as a tracer, radioactive salt (in which sodium-24 replaces some of the normal sodium-23) is injected into an artery in one part of the human body, and then a radiation detector is used to observe its arrival at another part of the body, thereby showing how well the blood is circulating. Detectors are so sensitive that only a very minute amount of the tracer sodium-24 is needed, a relatively few radioactively "tagged" molecules mixed with normal NaCl molecules. There are many special tricks of this sort, each using some appropriate chemical substance.

One of the most refined techniques is the location of a tumor by means of a beta-plus emitter. Positive electrons do not normally exist in nature because when one of them is created (in beta-plus emission) it quickly meets an ordinary negative electron and they literally annihilate each other (like the proverbial gingham dog and the calico cat that ate each other up). Their mass is turned into energy according to the famous Einstein formula $E = Mc^2$ (with M equal to $2m$, two electron masses). The important point here is that this energy is carried away by two gamma rays that travel in exactly opposite directions. If a tumor in the brain is suspected, the beta-plus emitter arsenic-74, for example, is injected into the blood in a chemical compound that has the property of seeking out the tumor. A pair of radiation counters on opposite sides of the brain is so connected as to count only when each receives a gamma ray. When the arsenic-74 reaches the tumor in great enough concentration, the gamma rays, traveling in opposite directions, will reach the counters. Thus when they count, one knows that the tumor is on the straight line between them. Several such pairs of counters can be used at once to locate the tumor accurately.

These are two of many diagnostic techniques using tracers in medicine. The other principal use of radioisotopes in medicine is therapy, or the treatment of disease.

We have seen that lethal damage to too many normal somatic cells can cause radiation sickness and death. But lethal damage to many cells in a malignant growth can be good. It can kill the malignancy often without the likelihood of spreading it by way of the blood stream, as often happens with surgery.

Perhaps the most dramatic therapeutic technique is the use of a moving beam of gamma rays to kill a localized malignant growth within the body. For this a movable cobalt-60 "gun" is used. This consists of a fairly small cylinder of cobalt-60 surrounded by a thick shield of heavy metal such as lead or tungsten. In one end of the shield there is a hole or gun barrel to let out a narrow beam of powerful gamma rays when a shutter is opened. Otherwise the shield confines the dangerous rays. The cobalt gun is carried by a heavy counter-weighted swivel mount in such a way that it is always pointing inward along a radius so as always to hit the center of the imaginary circle around which it swings. The patient is then placed so that his malignant growth is at the center (Fig. 5–4). The advantage of this arrangement

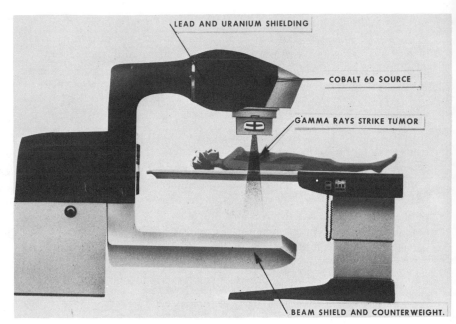

Fig. 5–4 A rotating mount for a cobalt-60 source to irradiate deep-lying cancers.

is of course that the malignant spot is always irradiated and receives a very heavy dose, lethal for the tissue, while the surrounding tissue is irradiated only a small part of the time and is not severely harmed. Ordinary light rays are bent when they enter a glass surface and can be focused by a lens to make a hot spot. Gamma rays are not bent, and this is about the only way one can focus them.

Cobalt-60 has a half life of five years, so the gun does not have to be loaded very often. It may contain a thousand curies or more, or as intense radioactivity as a kilogram of radium.

X-rays have long been used for such therapeutic purposes, but gamma rays are more penetrating and in some ways more convenient. X-ray pictures or x-ray shadows on a fluoroscope are also extensively used for diagnosis. Table 5–1 shows that almost all of the man-made radiation exposure in the United States is in the form of medical x-rays, and much more of it from diagnosis than therapy. (The diagnostic use has increased so rapidly that the 1971 figure is about twice the figure of 50 millirems given in the table for 1963.)

While it is important that care and moderation should be exercised in prescribing and administering nuclear radiation, it is thus even more important for x-rays.   X-ray fluoroscopy and photography have become such helpful tools for diagnosis that many doctors and particularly dentists use them routinely and with little thought of the cumulative physiological consequences, which are expected to show up only statistically in large populations exposed.   Modern equipment, properly filtered so as to expose the patient only to the rays that are photographically most effective and using careful development of fast film, gives typically about 10 millirems per chest x-ray (one-fifth of that listed in Table 5–1 as the annual average diagnostic exposure).   It is rare even in a good hospital that one can learn from the operator what the actual exposure is.   It is usual to insist that it is completely harmless.

The real culprit is x-ray fluoroscopy, which probably should be considered obsolete but is not.   A relatively high intensity of x-rays shining on a fluorescent screen is required to make it glow instantly and show shadows of bones and other structures.   There is sometimes important convenience in not having to wait for a photograph to be developed.   This is also economical, because a sensitive photographic film large enough to cover the whole chest, for example, is not cheap. For this reason, routine chest x-rays of large numbers of people are still commonly made, particularly in the traveling mobile units, by having a high enough intensity of x-rays to make a fluoroscopic screen glow and by then taking an ordinary camera picture of this on small (35 mm) film.   This method commonly exposes the patient to one rem, or twenty times what is listed in the table as the national annual average for diagnostic x-rays.   (The figure listed is probably low.)   Physical examinations for army recruits use this method and by inducing cancer probably kill several individuals among the hundreds of thousands examined.   The lives that the examinations save could also be saved by safer direct x-ray photography.

INDUSTRIAL USES OF RADIATION

The main industrial applications of radioactive isotopes may be divided into tracer techniques, radiography, control, and treatment.[6]   The tracer techniques are quite similar to those used in medicine; one inserts radioactively "tagged" atoms or molecules with appropriate chemical properties at one point of a system and watches them proceed in a host of normal atoms through some process until they arrive at a detector and indicate something about what happened along the way. A typical example is measuring the rate of wear of piston rings. A piston ring can be neutron-irradiated to change some fraction of the iron nuclei into radioactive iron-58. Then the ring is placed in a test engine and the rate at which radioactivity appears in the motor oil is observed.

Many commercial products are made in large sheets that pass rapidly between sets of rollers. Paper and sheet metal are examples. A radioisotope thickness gauge consists of a source of radioactivity, such as beta rays, on one side of the sheet and a detector of radiation on the other, so that the intensity observed getting through depends on the thickness of the sheet. This gauge may be used either for quality control, to monitor the thickness as the sheet rolls past, or actually to control the uniformity of the thickness by feeding the electric-current reading of the meter into machinery to control the position of the rollers, for example.

Gamma-ray radiography is much like x-ray photography with the difference that gamma rays are much more penetrating and can be used to detect flaws in thick metal objects, such as cannon barrels or welds in pressure vessels for nuclear reactors. A cobalt-60 gun similar to that used for medical therapy is typically used as a source, and the detection can be done either by photography or by electronic radiation detectors.

Radiation, mostly gamma radiation, is also used for treatment and processing of various industrial and agricultural materials. Beneficial properties of plastics and fibers are induced by rather weak radiation. Potatoes can be made not to sprout, and insects in stored grain can be killed. Sterilization can be accomplished even when materials are already packaged in a protective covering. Canned food so sterilized lasts well.

In the United States this rapidly expanding use of radioisotopes in medicine and industry grew up almost entirely with radioactive materials prepared at Oak Ridge National Laboratory, in Tennessee, mostly by neutron irradiation of materials inserted in a special reactor operated for this purpose. Materials from this source have been shipped abroad, but other countries also have governmental radio-isotope production facilities. In the United States the trend is for industry to take over some of the production. This is a challenging new field, a gratifying beneficial use of nuclear radiation. In this field also careful control and handling are needed to avoid excessive human exposures and contamination of the environment. Though the quantities of radioactivity are minute compared to the content of a nuclear power reactor, they are more closely handled by people.

REVIEW QUESTIONS AND PROBLEMS

1. What is an enzyme? a gene? a cell? a mutation?

2. When a cell undergoes "fission" (or "mitosis") how does each daughter cell receive the entire genetic information of the parent cell?

3. Why is a lethal mutation of a somatic cell ordinarily not harmful? What happens with very many such mutations? Why is a nonlethal mutation of a somatic cell ordinarily more serious than a lethal mutation?

4. Same questions as in (3) for a genetic cell.

5. Roughly how does the amount of radiation exposure to the population from medical uses compare with that from nuclear power plants?

6. In order of magnitude, what is considered to be the likelihood of damage to the child from use of pelvic x-rays on the mother?

7. What were the conditions under which two critics of radiation standards claimed that some 30,000 deaths per year might occur? Are those conditions likely to occur as the nuclear program is now progressing?

8. In what manner are most high-level radioactive wastes from nuclear power and bomb production now stored? What better method might be possible?

9. Explain how a brain tumor can be located by use of radioactivity.

10. How can gamma-rays be "focused" on malignant tissue?

11. How can the thickness of a plastic sheet be monitored by use of radioactivity?

CHAPTER 6

Control of Fissile Materials

The materials handled in the nuclear power industry are fraught with two kinds of danger, their radioactivity and their potentiality as materials for making nuclear bombs. In the previous chapter we have considered the importance of careful handling and sequestering of radioactive by-products of nuclear power production because of the biological hazards of their radioactivity. We now turn to the other aspect of the problem, to the fact that the fissile materials as used in some reactors and as produced by all reactors are the stuff that bombs are made of and could make the outbreak of nuclear war more likely if they should fall into irresponsible hands.

NUCLEAR PROLIFERATION

In the period immediately following 1945, at the close of World War II, the United States found itself in almost sole possession of the huge body of technical information that was then somewhat misleadingly referred to as "the secret of the atomic bomb" (as though it could be transferred as a single formula or blueprint). The United Kingdom, as a wartime collaborator in the nuclear effort, was privy to much of this information. "The secret" was jealously guarded with strict security procedures. Information about uranium was locked up so tight as to hamper some of the progress of science, even "pure" science. The army of occupation entering Japan found two cyclotron laboratories, the result of patient development by Japanese physicists for pure research, and stupidly destroyed the cyclotrons in spite of pleas from American scientists insisting that they were not atomic bombs. The vanquished nations, Japan and West Germany, were forbidden to participate in nuclear enterprises.

While the details of the construction of nuclear weapons remain highly secret to this day (and this is helpful for the purposes of avoiding proliferation), the attitude of extreme secrecy concerning nuclear matters peripheral to weapon technology was suddenly reversed in about 1954. Under the intriguing slogan "atoms for peace" a great show was made of releasing information on reactor technology and nuclear physics, and some conferences were held that were instrumental in bringing together scientists of East and West to their mutual benefit. Besides this, the United States Atomic Energy Commission was very active in promoting and financing the construction of nuclear reactors throughout the "free world," particularly small demonstration reactors for research and teaching purposes. Foreign students of reactor technology were brought to this country for instruction, enriched uranium was supplied for the reactors on a loan basis, and in general everything was done to promote interest in developing nuclear energy technology throughout most of the world.

This promotion accelerated the situation in which one had to worry about the possible entrance of many countries into the nuclear weapons race. The first demonstration reactors made very little plutonium, but the larger-power reactors that followed in several countries do make enough plutonium to constitute a nuclear weapons threat if not properly controlled. The International Atomic Energy Agency (known as IAEA), with headquarters at Vienna, was set up to help with the worldwide development of nuclear energy and to handle the problem of avoiding clandestine diversion to nuclear weapons purposes (Fig. 6–1). Impatient with the development of these international procedures, the United States entered instead into bilateral agreements with several dozen countries, arranging to supply them with enriched uranium provided that the spent fuel would be returned to the United States for processing so the plutonium could be extracted here and kept under our control. These bilateral agreements, avoiding the good offices of the IAEA, have left that organization with rather little of the type of activity that it was intended to perform. The nonproliferation treaty of 1970 belatedly called upon it to perform this function in a healthy fashion.

A modern power reactor contains a large quantity of uranium, enough to make each year plutonium for several dozen atomic bombs at a rate about half a gram per megawatt day of electric power. An

Fig. 6–1 International Atomic Energy Agency headquarters in Vienna.

atomic bomb is variously said to require about seven to ten kilograms of plutonium or ^{235}U. If the plutonium is to be used for bombs, the fuel should not remain long in the reactor, because the ^{239}Pu that is produced is subjected to the bombardment of neutrons which slowly change some of it to ^{240}Pu. This latter is a fast-neutron absorber and acts as a poison to inhibit the explosion of a bomb. It cannot be chemically separated from the useful isotope of plutonium. Thus if a reactor is run too long the plutonium in the spent fuel becomes unsuitable for bomb use. The 6 percent "burn-up" of ^{235}U in American commercial reactors is purposely small enough to avoid this difficulty. By requiring a long fuel cycle, a reactor can be prevented from making efficient bomb material. This possibility is not yet extensively used as a method of control. Plutonium thus "denatured," or poisoned for fast-neutron use, would still be useful in principle for slow-neutron reactors if the metallurgical and other difficulties of using plutonium could be overcome.

About ten countries now have large enough reactors, fuel-processing plants, and technical ability to develop their own atomic bombs if the political decision should be made to do so. Through a process of inspection it is possible to provide some assurance, but not very firm assurance, that plutonium is not being diverted to bomb use. Diversion of the fuel elements themselves would not be serious because of the lengthy processing required to remove the plutonium. The most sensitive spot in the fuel cycle is in the fuel-reprocessing plant where the plutonium is removed from the rest of the dissolved core, or in the subsequent transportation of the plutonium. In the wake of recent interest in a nonproliferation treaty, technical studies of ways to improve inspection are being carried out quite energetically.

In recent years both of the "nuclear giants" have become seriously interested in avoiding the diplomatic nightmare of having many nuclear nations, any one of which might start a nuclear war. Though it might start between small nations, such a war could easily escalate into one between the large nations as well, "catalytic war," as this grim prospect is called. Deterrence between two or a few powers has so far proved fairly stable, fortunately. But one worries about what would happen if there were many sources of nuclear trouble, some of them in politically unstable countries and countries not sufficiently technically advanced to provide adequate safeguards against military accidents.

During the 1950's there were three nuclear nations, the U.S., the U.S.S.R., and the U.K. (Great Britain). The U.K. had collaborated with the U.S. in the wartime development of the A-bomb and, because of U.S. secrecy, largely had to go it alone in developing its H-bomb— which it never deployed in large numbers comparable to those of the U.S. and the U.S.S.R. One spoke of the problem of proliferation as the "fourth-nation problem." Despite her generally friendly relations with the United States, France also had to go it alone in her development of nuclear weapons, so consistent was the policy of the United States to avoid contributing to the spread of weapons information and thus to discourage proliferation. China found herself in a similar situation vis-à-vis the U.S.S.R. Relations between them suddenly cooled in 1958 after the U.S.S.R. had supplied China with reactor materials and technology, so China had to go it alone in isotope separation and bomb development. The timetable of achieving nuclear bombs was as follows:

	A-bomb	H-bomb
U.S.A.	1945	1952
U.S.S.R.	1949	1952
England	(1945)	1957
France	1960	1968
China	1964	1967

The fact that France, an advanced industrial nation, required about a dozen years of intensive and expensive effort to get past the A-bomb stage and get a separation plant working in order to make an H-bomb indicates how important the control of nuclear materials might be in slowing down further proliferation to other countries. However, the experience of China, an enormous but industrially underdeveloped nation, indicates that many nations can arrive at this stage if they have the motivation to go to the expense and effort. It is felt by most nations that five nuclear nations are already too many for world stability and that a larger number would further increase the likelihood of nuclear war. Attempts are being made to keep the number from growing.

THE NONPROLIFERATION TREATY

Negotiations have led to the writing and signing of the Nonproliferation Treaty (NPT). This treaty was first promoted by the nuclear powers in 1968 and requires less restraint on their part than on the part of the nonnuclear nations. It is, however, in the course of being accepted by most of them. While the treaty requires each of the nonnuclear nations to give up its sovereign right to make nuclear weapons, this may be seen by each of them as in its best interests because it is safer if neither it nor its neighbor has nuclear weapons than if both do, as clearly illustrated in the case of Israel and Egypt, for example.

The treaty contains three principal provisions:

1. The nonnuclear nations will not accept nuclear weapon materials or nuclear weapons from abroad and will not develop them. As proof of good faith they will accept IAEA inspection (or in some cases, temporarily, equivalent inspection by the more local European organization Euratom).

2. The nuclear nations will not supply nonnuclear nations with such items.

3. The great nuclear powers will enter into sincere negotiations to limit the nuclear arms race.

For the other, less important, provisions, see Appendix 18.

It was provided in the treaty that it would go in force when signed by the major nuclear powers (China excluded) and when it was signed and ratified by a total of 42 nations. These requirements were satisfied in March of 1970, and as of that date the treaty went into force binding those nations which had signed. Serious exceptions were India and Israel.

It must be admitted that such a treaty by itself will not stop with certainty the proliferation of nuclear weapons. The foreseeable expansion of nuclear power plants will make the problem increasingly difficult. However, the treaty will make it more difficult for a nation to decide to "go nuclear" or for some criminal organization to acquire nuclear weapons. This is a useful step in taming the arms race, but it must be followed by others if real stability is to be attained. A complete test ban, forbidding not only atmospheric tests but also those underground, would be an important and useful supplement.

DANGER OF DIVERSION OF FISSILE MATERIALS

As potential nuclear explosives, fissile materials are so dangerous when combined with a considerable amount of technical capability, that it is extremely important they should be kept out of irresponsible hands. Yet these materials are rapidly becoming so commonplace, not only in the armed forces but outside them, as to give cause for serious concern. While the cause of nonproliferation was given a boost by the signing of the Nonproliferation Treaty in 1970, it was set back by the legislation in 1967 authorizing private ownership and industrial processing of these materials.

It is remarkable how quickly the political world has become calloused to the dangers of diversion of nuclear materials. In the early 1950's the AEC Chairman, David Lilienthal, was publicly rebuked by Senator Hickenlooper for "incredible mismanagement" when five grams of ^{235}U could not be accounted for at a national laboratory. Up through most of the 1960's fissile materials were the property of the United States Government (the AEC) and were leased to the utility companies operating power reactors, to be returned to the AEC for reprocessing. Considerable caution was exercised lest some be diverted, to the extent that a responsible courier accompanied important shipments. Since 1967 all that has changed. Shipments of enriched uranium are carried by ordinary trucking companies at the same rates and with the same amount of caution as a shipment of shoes—much less than a shipment of furs. It is now a common occurrence for significant amounts of fissile materials to go astray in shipment. As far as is known, these have eventually all been recovered, but the fact that they go astray suggests how easy criminal diversion could be. In an example cited as typical by Congressman Craig Hosmer in May, 1970, as an impromptu part of the speech[2] noted in Appendix 15, 80 kilograms of highly enriched ^{235}U, shipped from one commercial firm in New York State to another in Ohio, failed to arrive at the Columbus airport and after a week of frantic search was found somewhere in Texas.

When enriched uranium-235 or plutonium is prepared and handled for military use, it is classified as such and is transported, quite properly, with great care to avoid diversion. Here the expense of a special courier comes from the military budget, and there is no artificial demand for economy. Careful handling is considered to be an essential element of national security. But when the same type of materials are shipped from one commercial concern to another in connection with nuclear power production, the less cautious handling leaves opportunities for highjacking along the way. The contrast no doubt arises from the influence the nuclear power interests have on Congress in their efforts to make nuclear energy seem competitive with fossil fuels. It has been demonstrated that it is easy for an unauthorized driver to pick up in a truck-transfer station a truck loaded with nuclear materials and drive off with it and even return it to where he found it parked without being detected. It seems imperative that this degree of economical laxness in the nuclear energy program be rectified. (See Appendix 15, Shapley, Hosmer.)

ACCOUNTABILITY DURING PRODUCTION

There are other possible leaks of fissile material that are more difficult to plug. Government and international agencies are making studies of accountability and the possibilities of diversion from various parts of the fuel-processing cycle. The unloading of the fuel from the reactor and its transportation to the fuel-processing plant have their radio-active hazards, but they are not dangerous points of diversion because the fuel is usually not in a form suitable for bombs and cannot easily be made so. The most sensitive points are those parts of the fuel-processing plant where the plutonium is separated out, handled, and eventually fabricated into fuel assemblies for trial use in reactors if it does not go into storage or bombs. In the chemical vats used in the separation process, it is hard to be sure within better than perhaps five percent just what the plutonium content is. If a worker could manage to tap off an amount less than that margin allows, he might not be detected by the materials accounting. A single big water-moderated reactor, not primarily a breeder, makes enough plutonium for several hundred atomic bombs a year (at roughly half a gram per megawatt-day thermal). A fuel-reprocessing plant handles the output of several reactors. A worker taking out a little every day would have to pilfer only a very small fraction of the plutonium going through the plant to get enough for a bomb in a year. This kind of activity might feed into a dangerous black market in these valuable and dangerous materials.

This is the type of problem that the International Atomic Energy Agency inspectorate is getting ready to handle,[1] and there is a group in the United States trying to help explore the technical aspects of the problem, such as the possibility of setting up continuous monitoring

Fig. 6–2 IAEA inspector observing remote-control handling of radioactive material at the Nuclear Fuel Services plant in West Valley, New York.

devices at key points in the flow pattern.   Unfortunately, the IAEA budget for this type of activity is rather low, about a million dollars annually, small even compared to the AEC budget of about six million dollars for its "safeguards" division. It is a program that must grow rapidly with the industry. At present the IAEA activities consist largely of inspecting the methods and accounts of various plants (Fig. 6-2) and seem quite inadequate to the larger task. It seems as though the control mechanisms are trailing, trying desperately to catch up to the need as the nuclear industry forges ahead to make more and more of the materials that need to be controlled, regardless.

Already the problem of diversion is serious. As we look into the future, if the world goes the way of a "plutonium economy" based on breeder reactors, its magnitude becomes staggering indeed. This is perhaps the most urgent basis for the hope that we may be rescued from a future of fission by the advent of another source of power.

The prospective economy of breeder reactors is based on the expectation that they will produce not only power and enough new fissile material to replace that which they consume, but will also have a substantial surplus to sell to the makers of new reactors. It is thus based on the presumption of a nuclear industry experiencing continual growth—even exponential growth. Sometime not very far off this growth must exhaust the resources of the global ecology, and the expansion must taper off. Then the momentum of industry seeking new outlets in a gutted market could further increase the threat of clandestine bomb-making.

REVIEW QUESTIONS AND PROBLEMS

1. How many nations have nuclear weapons? How many of them have H-bombs?

2. Which nation first achieved an A-bomb? An H-bomb?

3. What are the main provisions of the Nonproliferation Treaty?

4. Is a special courier required to accompany shipments of commercial fissile material in the United States.

5. What part of the cycle through which the fuel of a reactor goes is particularly vulnerable to diversion for illicit purposes?

CHAPTER 7 Other Possible Power Sources and Future Needs

FUSION: THE QUEST FOR CONTROLLED THERMONUCLEAR POWER
The various difficulties associated with production of electric power from fission—particularly the pollution hazards and the prospect of early exhaustion of the easily recoverable supplies of ^{235}U—have spurred research in the hope of developing another source of nuclear energy, fusion. It is fusion that powers the sun and explodes in the H-bomb, but the challenge is to contain and tame it in the laboratory and ultimately in the power plant.

Figure 7–1 is a repetition of Fig. 3–14. It reminds us that the internal energies of nuclei vary in such a way that the energy of nuclei *per nucleon* is least for the intermediate-mass nuclei, in the neighborhood of mass numbers $A = 70$ to 150, the curve rising sharply toward the very-light-nucleus end and gradually toward the heavy-nucleus end. This means that it is possible to decrease the internal energy and release the excess energy for external work in a process moving from either end toward the middle of the range of A. Fission moves from the heavy end toward the middle by splitting an almost unstable heavy nucleus into two intermediate-mass nuclei and releasing energy. Fusion moves from the light end a short way toward the middle, down the steepest part of the curve, by synthesizing two very light nuclei to make a somewhat heavier nucleus, and releases energy. If we want to judge what new miracles science is apt to provide in the future, it is important to appreciate that these two processes exhaust the possibilities of gaining large amounts of power from rearrangement of nuclear matter, coming in from either end of the curve toward the low-energy middle. Nuclei can be split at the very heavy end because there

162

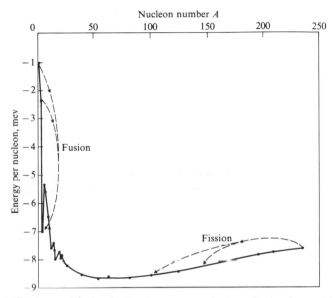

Fig. 7–1 Fusion and fission both release energy by transferring from a high end toward the low middle of the energy-per-nucleon curve.

they are on the verge of being unstable. They can be pushed together, with difficulty, by use of high temperatures, at the very light end because there the electric charges are sufficiently small. People who don't understand these facts are apt to expect science to go on pulling new energy-producing miracles out of this hat.

The small slope of the curve toward the high-A end might give the impression that not much energy is to be gained there until one remembers that the curve gives energy *per nucleon*. There are many nucleons involved in fission, so the energy release is great. There is only about 1 Mev difference between the right end and the middle, but since there are over 200 nucleons involved, about 200 Mev of energy is released. At the other end the curve is steep and the energy release per nucleon is great, but there are many fewer nucleons involved. Thus in a fusion process the energy release is only about a tenth as great as in fission.

We have learned that the curve slopes upward at the large-A end because the electric repulsion between protons becomes dominant there, and indeed fission derives its energy from that repulsion. The curve slopes downward at the small-A end because there the short-range attraction is dominant and is used more efficiently when more nucleons are put together, in spite of the electric repulsion between the very few protons.

There is, however, enough electric repulsion between light nuclei that they must be thrown together very energetically to make them touch each other and amalgamate or fuse with one another. This is the great barrier to an easy and practicable controlled fusion process for the release of energy. It is not difficult to do on an experimental basis with an accelerator. About $\frac{1}{10}$ Mev is sufficient for such experiments. But an accelerator consumes enormously much more power than is released in a few such reactions. So electric repulsion is both what makes fission go and makes fusion power hard to attain.

The trick in a thermonuclear reaction is to give many light nuclei some kinetic energy of thermal agitation by making them very hot. Thus one has very many nuclei colliding with very many others in the hope of obtaining many individual fusion reactions to supply the energy to continue the heating. The thermal agitation of the molecules or atoms or ions in a hot gas is a helter-skelter business. In a gas at a given temperature, most of the atoms have kinetic energies fairly near the average. Very few have almost zero and a few have kinetic energies way above average, as indicated in the curve of Fig. 7–2. The whole curve shifts toward higher energies as the temperature increases (and in direct proportion: $KE_{av} = \text{const} \times T$). The few ions out in the "high-energy tail" are the ones with enough energy to cause fusion when the temperature is high enough. Thus it is important to get the temperature very high, to push the high-energy tail out as far as possible, to have as many ions as possible with enough energy to cause fusion, and to attain a high yield of fusion energy.

The temperatures needed are of the order of magnitude 10^5 degrees —far too hot for any solid container to handle. It is found possible instead to confine a very hot low-pressure gas by means of magnetic effects.

A very hot gas at fairly low pressure is known as a plasma. In it most of the atoms are ions, positively charged because of loss of some of their electrons. The positive ions and the electrons bounce around together in such a way that the average charge in any fairly small volume remains zero (for if there were a net positive charge in some

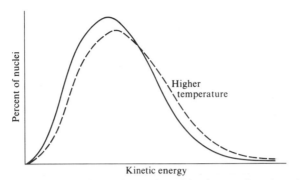

Fig. 7–2 Fraction of particles having various kinetic energies when the plasma has a given temperature.

region, it would attract electrons to it). In this respect a plasma is something like a metal, though in the metal the closely packed positive ions are fixed in place and electrons can swim around through them. The flow of electrons means conduction of electricity; both metals and plasma are electric conductors.

When a metal wire carries a current of electrons past the poles of a magnet, there is a sideways force on the wire. That is what makes electric motors go, as we have seen. Conversely, when a wire is pushed sideways past the poles of a magnet, an electric force and (if the ends are suitably connected) an electric current are generated in the wire. That is what makes electric generators function. Similarly, when an electron in free space flies past the poles of a magnet, it feels a sideways force and is deflected sideways. Such a sideways force makes the ions go around a circular path in a cyclotron, or makes it possible to separate isotopes by deflecting light ions more than heavier ions. And, to get back to a plasma, that is what makes it possible to confine a plasma by use of magnetic fields.

A magnetic field is a condition that exists between the poles of a

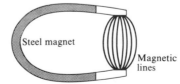

Fig. 7–3 Magnetic lines between the poles of a magnet.

magnet. It can be represented by bulging lines drawn between the poles as sketched in Fig. 7–3. The lines tell the direction a little compass needle there would point. The force on a moving ion is at right angles to the direction of these lines. There is a similar magnetic condition around a wire carrying electric current, represented by similar lines circling around the wire. By use of many current-carrying wires suitably arranged it is possible to shape these lines into a "magnetic bottle" around a hot plasma, to hold it in place (Fig. 7–4). Whenever a fast-moving electron or ion comes to the edge of the plasma, as if attempting to escape, it suddenly finds itself in the magnetic region and is sharply deflected around in a semicircle and is sent back into the plasma. That is how the magnetic field confines the plasma without any loss of kinetic energy, and thus without cooling off the plasma as collision with a material wall would do.

One trouble with a magnetic bottle such as sketched in Fig. 7–4 is that it leaks a little bit at the ends. To avoid this a long "bottle" is sometimes bent around like a doughnut so there are no ends.[1]

The isotopes of hydrogen are the most suitable ions for thermonuclear reactions because they are repelled by the weakest electric force, since they each contain the electric charge e of only one proton making them repel each other with the simple inverse square force e^2/r.

Two heavier nuclei a distance r apart, one with Z_1 protons and the other Z_2 protons, have electric charges $Q_1 = Z_1 e$ and $Q_2 = Z_2 e$ and push each other apart with the larger force

$$Q_1 Q_2 / r^2 = Z_1 Z_2 e^2 / r^2$$

as described in Appendix 4, page 271. This would require much higher temperatures to make them come together for the larger values of $Z_1 Z_2$. For example, even two lithium nuclei, with merely $Z_1 = Z_2 = 3$ and thus with $Z_1 Z_2 = 9$, repel each other nine times as strongly as do two hydrogen nuclei, such as D + D or D + T with $Z_1 = Z_2 = 1$ and $Z_1 Z_2 = 1$.

One important nuclear reaction is

$$D + D \rightarrow T + p + 4 \text{ Mev}.$$

This is reaction 2 in Table 3–3. (Reaction 3 there is also important, particularly in the initial heating up, to get the process started.)

In some of the experiments a region of plasma inside a magnetic

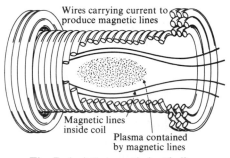

Wires carrying current to
produce magnetic lines

Magnetic lines
inside coil

Plasma contained
by magnetic lines

Fig. 7-4 A "magnetic bottle."

bottle is first heated up as much as possible by some external means, such as bombardment by a powerful "electron gun." Then the plasma is suddenly very much compressed by increasing the current in the wires so as to pinch in the sides of the bottle. Compressing the gas of course heats it further, and it has been found possible in this way to get the thermonuclear reaction started. The hope is that the energy release from each nuclear reaction, 4 Mev in the case just cited, will further heat the gas and help sustain a high temperature without more pinching.

The trouble that has long plagued research in this field has been instability at the edge of the plasma. In the conflict between the outward pressure of the plasma and the inward magnetic pressure there, the magnetic lines tend to get all tangled up. Back in the mid-fifties, when this trouble was first encountered, enthusiastic research workers thought that the trouble could be overcome and thermonuclear power produced within perhaps five years. Much progress has been made in the meantime, and some workers are again optimistic about obtaining a laboratory version of a sustained thermonuclear reaction. Even if that could be attained, the path from there to practical generation of power would still be a long one fraught with other uncertainties.

What has been done is to produce a thermonuclear reaction in a small volume at a low pressure for a small fraction of a second.[1] The laboratory goal is to produce it in a large volume at a higher pressure continuously. In the long course of research, progress has been measured in terms of the product $P =$ (volume) $\times$ (pressure) $\times$ (time reaction lasts). The product has been creeping up from something like 10^{-6} to 10^{-2} of what is needed, the last factor of 100 or so coming in a recent Soviet advance with an experimental model known as Tokamak.

It is most gratifying that this aspect of applied science has shared with pure science a spirit of friendly and entirely cooperative rivalry between Soviet and western scientists. Complete exchange of information, with frequent conferences and visits to one another's laboratories has meant a rapid pooling of all the best ideas on a worldwide basis and healthy stimulation of progress. To those who feel directly the benefits of this sort of trustful collaboration, it is frustrating to observe, in the more immediately life-or-death matter of military competition, the inability of nations to avoid artificial distrust and through mutual trust to find benefits in terms of greater safety and material well-being.

The hope for fusion is to produce power using the deuterium of the sea as fuel. Deuterium is a rare isotope, one part in about 20,000 of the hydrogen in seawater, but the seas are large and the supply practically inexhaustible. Among the very many questions to be faced is whether, if power could be produced, it would be more than enough to prepare the deuterium fuel consumed.

One advantage of thermonuclear power should be that it should produce considerably less radioactive pollution than does fission power. In particular, there would not be the problem of disposing of high-level wastes of fission products. Of the low-level radioactivity that is released from fission power plants, part is normally fission products, not only gases like krypton-89 but also other products that leak or diffuse through the fuel cladding. But a considerable part of what is flushed into the river comes from neutron activation of reactor materials, and this might be just as serious in a fusion power source. Indeed, a fusion source may make use of neutron irradiation of lithium outside the plasma as an important part of its energy production.

If it could be achieved, thermonuclear power would be a wonderful boon to mankind. But it is a chicken that should not be counted before it is hatched. The technical problems are so difficult, it would seem almost a miracle if it comes to pass. In the political world there is a tendency to avoid taking measures to forestall future troubles in confidence that science will somehow provide, as it has on so many occasions in the past. In this instance there is a tendency not to worry about the fact that we are using up our uranium supplies rapidly when we know how to use only the ^{235}U component, without awaiting a more efficient technology, because of a feeling that by the time the uranium is used up at this rate we will have fusion power. This is unjustified optimism. There are many serious obstacles in the way, and no assurance at all that science will surmount them. It is not a sure thing, but perhaps with some odds a good bet. Some of those working on the project are optimistic enough to believe that there is more than a 50–50 chance of attaining a laboratory demonstration of practicability by 1985. This would mean a demonstration giving out more power than is consumed in maintaining the magnetic fields, for example. If that should be attained, it might be another 15 years before the formidable engineering obstacles could be overcome to make a practical electric power plant.

Thus fusion power must be looked on as an exciting scientific and technical challenge highly deserving of generous funding and encouragement of research and development, but definitely not to be counted on in planning the management of future power needs.

If a thermonuclear electric plant could be attained, it would have the advantages (1) that the prospective fuel supply is almost limitless (if it can be made to operate mainly on deuterium), (2) that it would make much less radioactivity, and in particular none of the long-lived fission products that have to be carefully isolated from the biosphere "in perpetuity," (3) there would not be the large inventory of fission products in the plant that might escape in an accident, and (4) it would operate at a higher temperature (T_{in}) and thus do less heating of the cooling water per electric kilowatt. The radioactivity, while less than with fission, could still be troublesome, both because of the tritium produced and because of the radioactivity induced in the materials of the machine.

With such enormous potential importance for the future, it would seem that a great deal of research effort should be going into fusion. A considerable amount is, yet the United States budget for fusion research is only about a quarter that for fission research (about $25 million compared to $100 million) in spite of the fact that fission technology is already well along. The Soviets are putting about four times as much effort on it as we are. Fusion is a cooperative effort with a constructive aim, and it would seem more fitting that we should be pulling our oar equally hard here rather than putting so much emphasis on staying ahead in the arms race.

SOLAR ENERGY

A discussion of solar energy belongs in a treatment of nuclear energy on two counts. First, solar energy is nuclear energy, for the sun is a thermonuclear furnace. It has the advantage of gravity and of great thicknesses of matter to confine its very dense sort of plasma and does not need a magnetic bottle. It feeds on not only the D + D and D + T reactions so important to earth-bound thermonuclear hopes but also on various other reactions between light nuclei, particularly those up to charge number $Z = 7$.

But second, the possibility of harnessing solar energy on earth is the most hopeful alternative to nuclear energy (including fusion) to support the industrial civilization of the future. Its prospects must be weighed to judge whether we need to run the long-term risks of really plentiful nuclear power.

It may seem strange to talk about harnessing familiar sunlight for that—it has been shining on earth so long, and nobody has come close to doing anything of the sort. Nature has done it with the chlorophyll of green leaves and has stored some of the product in fossil fuels, but on a continuing basis this supplies little more than an inadequate supply of wood to burn and the food that keeps animal muscles working, including our own.

The energy of sunlight falling on earth is about 1 kilowatt per square meter. (To appreciate this, think of a familiar 100-watt light bulb at the center of a paper sphere of $\frac{1}{10}$ meter (4-inch) radius. The inner surface of the sphere is about $\frac{1}{10}$ square meter ($4\pi \times 0.1^2$). All of the $\frac{1}{10}$ kilowatt from the bulb falls on $\frac{1}{10}$ square meter. A considerable part of the 100 watts comes off as light, some as heat. A piece of paper 4 inches away from a 100-watt bulb is illuminated almost as brightly as by sunlight.) Converted to larger figures, this means that sunlight falling on 1 square kilometer (0.4 square mile) is 1000 Mw, about the power of the largest nuclear reactor now being built. If we could utilize the energy falling on a square mile with 40 percent efficiency, this would be equivalent to one large nuclear power station. If it seems fantastic to think of covering a square mile with rather intricate machinery, one should think of the care and expense that must be lavished on a nuclear reactor.

There are several ideas in circulation for harnessing solar power.[8] One coming from people interested in continued funding of the space program proposes to unfold square kilometers of delicate receptor in space and then to beam the power on another wavelength down to earth. Such phantasmagoria, if indeed they be that, should not be allowed to discredit serious research in down-to-earth harnessing of solar power. One down-to-earth idea involves developing cheap lenses that do not need to be kept very clean and to cover an area of desert with them, aimed in banks and each focused on a small furnace in which the high temperature achieved would induce a chemical trans-formation to make fuel. It may be a long way from the dreaming to the doing, but it seems a shame that some ideas in this field have not been really vigorously pushed with even a small fraction of the kind of effort that has gone into nuclear energy.[8]

As has been indicated in the introduction, the other sources of power, geothermal, tidal, hydro, and in practice probably also wind, are not plentiful enough to supply large parts of the future demand. They can make important limited contributions in special circum-stances, as can local solar heating of homes by way of heated water tanks.

Thus it appears that in the long term, on into the centuries, the continuation of industrial civilization will require either fission power (via some kind of breeder), fusion power, or solar power.

FUTURE POWER DEMAND AND NEED

We have seen in Chapter 1 that the overall consumption of power in the United States has been doubling about every 24 years in its expo-nential rise. The consumption of electric power has been doubling every 9 or 10 years, and the electric industry, among others, firmly expects it to go on doing so. The population increases about 10 percent in 10 years so this corresponds to a per capita consumption of electricity doubling every 10 or 11 years—roughly each decade. In most dis-cussions of future power demand or need, a continuation of the expo-nential rise at this rate is taken for granted. AEC spokesmen and the Federal Power Commission reflect this expectation (Fig. 7–5). One dissenting participant in a recent power conference where this opinion prevailed referred to it as the "exponential idiocy syndrome." It is clear that these exponentials cannot go on rising forever. The question is how soon they should start tapering off.

Power demand and power need are not necessarily the same thing.

Fig. 7–5 Cover of a report indicating expectation of continued exponential growth of electric power consumption.

Demand is artificially stimulated. The same utility companies that are pointing to the doubling each decade as necessitating their "going nuclear" are together spending over $300 million a year advertising the use of electric power; this is almost ten times what they spend on research while there is need for more research on cleaning up fossil fuel plants.

The reason for the increasing demand is largely economic. Ever since the arrival of the first Europeans on these shores, Americans have been accustomed to a rapidly expanding economy. Expansion has meant prosperity. Now that the geographic frontier has practically vanished, technological frontiers take its place. The standard of living of the majority of Americans, at least as measured in terms of the convenience of using goods and power, has been increasing to the mutual benefit of the producer and the consumer. Economists hardly know how to think in other terms than a rapidly expanding gross national product. It may be that a large part of our demand for continued rapid growth is not because we need to grow but because we don't know how else to behave, because institutions tend to perpetuate themselves and many of ours base their ideas of corporate profits on assumed growth.

Not all Americans enjoy the high standard of living of the comfortable majority. Some growth is justified to spread it farther, but this accounts for only a small part of the growth of our gross national product. Those living on low, often publicly supplied, incomes in cities consume more goods and power than their counterparts not yet displaced from rural settings. There are almost no Americans living on as low a material standard and consuming as little as many of the poor of other parts of the world. Advancing technology, of which electric power generation is a part, can and should be used to better the lot of mankind. The life of American poor should be bettered, but this in itself requires only a relatively small increase in power consumption. In a world view the problem of bettering the lot of mankind lies mainly elsewhere.

We are often reminded that we Americans constitute less than 6 percent of the world's people and consume 35 or 40 percent of the world's goods and power. Our $1\frac{1}{2}$ million-megawatt annual consumption of electricity (as of 1971) is more than a third that of the world. In this perspective we should ask how much more we need, on how rapidly rising a curve and when the curve should taper off to a steady level.[8]

Overall use of power in the United States is divided roughly as follows:

Nonelectric	
Transportation	1/4
Industry	1/4
Heating, air conditioning	1/4
Electric	
Industrial	1/8
Domestic	1/8

The electric sector is growing faster than the total, as we have seen (doubling time 10 years as compared to 24 years for the total). The industrial half of it grows because more and more industrial processes (such as the making of metal) are being transferred to electrical methods that are convenient and economical as long as electricity is kept cheap. Iron and steel are produced less in coke-burning furnaces, more in electric furnaces. As basic metals become harder to find and mine, more power is needed for metal production. Making aluminum takes lots of electric power, and because such power is cheap, aluminum is increasingly used for inexpensive articles such as cold-drink cans. A government project made power especially cheap in the Tennessee Valley, and industry makes aluminum there. The Tennessee Valley Authority is now turning to nuclear power for its continued growth (beyond its hydroelectric base) because of an artificial and hopefully temporary shortage of coal. This was caused partly by closing mines because of unjustified expectations of nuclear plants elsewhere.

The domestic half of the electric sector grows because space heating of houses, for example, is increasingly being done electrically with resistor units that are inefficient but cheap to install, more streets are being brightly lighted as an attempted technological answer to the social problem of increasing crime rates, and people are using more and more electric appliances. These are examples of the growing demand that is increasing the still relatively small electric sector. Reduction of the rate of growth of the electric power supply would not need to imply curtailing the use of all home electric appliances that are so very convenient in daily life, for most of these use much less electricity than

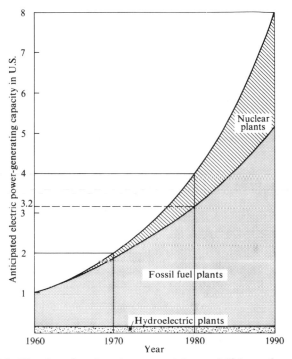

Fig. 7-6 Modification of projected exponential growth if the nuclear component were missing.

is consumed by electric heating of houses in northern climates, which most needs to be limited, and next to that, air conditioners and water heaters.

The nuclear share reached 2 percent of the electric generating capacity in 1971 and is anticipated to reach 20 percent by 1980 if the present AEC-licensed industrial program continues. This implies growth rates for fossil fuel plants and nuclear plants as indicated in Fig. 7-6. If we were to eliminate all the nuclear plants and depend on just the anticipated growth of fossil fuel plants, the electric generating capacity would increase 60 percent in a decade rather than 100 percent. The nuclear program of the next decade is expected to close that 40 percent margin. When we assess the importance of having the nuclear program, we must ask how essential is that marginal growth. Why should we not grow up the lower edge of the shaded portion of the figure, rather than the upper edge? Would not an economy growing 60 percent in a decade, or even 40 or 20 percent, be a healthily growing economy? Will it not perhaps represent an even more healthily growing economy as an approach to the approximately zero growth that will ultimately be necessary in man's long-term adjustment to his finite resources?

If one can achieve a viewpoint outside of the compelling competitiveness of commercial life, that is a relevant perspective in which to judge our real needs for power here in the United States, already a land of plenty.[5, 7]

The world perspective is different. The population growth of India, South America, and elsewhere, while dangerously rapid, is already limited in part by disease arising from starvation. The prospect of a world population overtaking the food supply is frightening. To a certain extent energy can produce food, and energy needs should be judged in that light. The agro-industrial complex, enthusiastically promoted by Alvin Weinberg,[3][9] could bring help to local spots in the underdeveloped world with essentially present techniques soon.

The more crucial need is for technology to provide a power source in the longer term, perhaps early in the next century when world population will be more insistently pressing the food supply.[3,4] By then it will be important to answer the question: Which of the possible sources is the one that should be exploited, solar, fusion, or fission? If fission, a breeder, of course, but which breeder? This question seems to stress how important it is to spare no effort now in research on all the hopeful types of possible future power sources rather than to permit competitive pressures to turn all our efforts to achieving short-term and local growth that from a larger perspective may be unneeded and harmful.

With the possibility in mind that solar or fusion power might in the future make it unnecessary to go on with the risks and disposal problems of breeder reactors, we should reassess this whole problem of power needs and uses. We should decide whether it is not better to go on depending almost entirely on cleaned-up fossil fuel consumption[16] for another decade or two until technology advances enough to indicate what the best future long-term methods for producing power will be. In the meantime let us recognize that all power pollutes,[2] so that demands for electric power should not be artificially stimulated.

REVIEW QUESTIONS AND PROBLEMS

1. What is the basis for the belief that after fission and fusion there will be no more fundamentally different ways discovered to develop large amounts of nuclear power?

2. In what way does high temperature promote a thermonuclear reaction?

3. Why cannot glass or ceramic or metal walls be used to contain thermonuclear plasma?

4. What is a "magnetic bottle"?

5. Would fusion power production, if feasible, produce radioactive wastes?

6. How long has it been taking for the overall consumption of power in this country to double? Of electric power?

7. About what fraction of the world's goods and power are consumed in the United States?

8. What fraction of the electric power production of the United States is expected to be nuclear in 1980? What fraction of the electric power of the United States was nuclear power in 1971? About what fraction of the total power consumption of the United States was that?

9. What are some of the factors making the demand for electricity increase so fast?

10. What is the electric equivalent of the intensity of sunshine (that is, in kilowatts per square meter or per square kilometer, e.g.)?

11. In your opinion and/or from what is said in the text, do you believe the probability of man's achieving practical electric power from fusion in the next century is less than 5 percent, between 5 and 95 percent, or more than that?

CHAPTER 8 Nuclear Explosives

THE A-BOMB

An atomic bomb or A-bomb gets its energy from a very sudden fission chain reaction. In principle it is similar to the more moderate chain reaction in a power reactor, but it depends on prompt neutrons and mainly on fast neutrons. The neutrons emitted by the flying fission fragments may bounce several times off uranium nuclei before they succeed in causing the next fission, but these encounters with heavy nuclei don't slow them down much. The fission cross section of ^{235}U (or ^{239}Pu) is smaller for fast neutrons than for slow neutrons, but sufficiently large (and considerably larger than that of ^{238}U) to make an efficient A-bomb. Almost pure ^{235}U or ^{239}Pu is needed.

We have seen that the intensity of a chain reaction grows if the fission ratio—the average number of neutrons from one fission causing another fission—is greater than one. The intensity grows rapidly if this ratio is considerably more than one. For a certain mixture of atoms, say 90 percent ^{235}U and 10 percent ^{238}U for example, assembled as a solid sphere with a certain density, there is a critical size of the sphere such that the fission ratio is equal to one. In the case of a bomb, this is the amount of material that is on the verge of exploding, and is known as a "critical mass" (or, in the vernacular, a "crit"). For a sphere of the material smaller than that, the fissions on the average take place too close to the surface and too many neutrons are lost from the surface to permit a sustained chain reaction.

Conceptually, the simplest type of A-bomb is the gun assembly (Fig. 8–1). The sphere of uranium may be prepared in two parts, rather as if a fat core were cut out of an apple. Each of the two parts is less than a critical mass, so they do not explode if they are rather far apart.

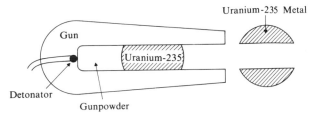

Fig. 8–1   Gun-type assembly of an atom bomb.

The trick is to put them together very rapidly. If they were put together slowly, the chain reaction would proceed far enough to generate enough energy to blow them apart, but just barely and it then would be a gentle explosion, a "fizzle." The nearly cylindrical inner part of the uranium is placed in a gun barrel to be shot, very much like an artillery shell, into a hole in the outer part of the sphere. At the instant when the sphere is completely assembled, the chain reaction is most intense and an enormous amount of energy is released, vaporizing the whole apparatus and sending out an extremely intense shock wave into the surrounding air. Most of the energy is created directly as kinetic energy of the flying fragments, and thus thermal energy, but about 10 percent is in the form of gamma radiation and more slowly emitted beta radiation.

During World War II, two quite different types of bombs were made—the gun type and the implosion bomb. Both were developed to be sure that at least one would work, and it turned out that both did. Only a single A-bomb of the gun type was made and was used at Hiroshima. It used ^{235}U. Recently the gun technique has also been employed to make fission-type nuclear explosives in the shape of long thin cylinders to be let down oil wells, as a part of Project Plowshare (discussed at the end of Chapter 9).

Modern A-bombs are of the implosion type. In this type the fissile material is compressed. Under ordinary pressure a certain kind of metal has a definite density, or mass per unit volume. The ordinary density of uranium is 19 g/cc. (That is very dense. By way of comparison, we ordinarily think of lead as a very heavy metal but it has a density of only about 11 g/cc.) But under very high pressure metals can be compressed to higher density. For a given amount of fissionable material the critical mass depends on the density. Instead of suddenly adding more material at normal density to make a mass more than the critical mass, as in the gun type bomb, it is thus possible to start with a given amount of material that is not quite a critical mass at normal density and suddenly increase its density to make it more than a critical mass at the new density. The crucial question is what fraction of the neutrons escape from the surface.

Just as in the case of a reactor, the critical mass of an A-bomb can be understood in terms of the assembly of mouse traps discussed in Chapter 4. There the "imitation neutrons" come back to hit the fission cross section (the trigger of the mouse trap) in the same plane from which they started. They are more apt to miss if the density of traps is reduced by spreading the traps farther apart. In the bomb the real neutrons either hit or miss the fission cross section of uranium atoms in successive planes which they pass as they move out in all directions from where they started. The neutrons are more apt to miss and escape from the edge of the uranium core if the density is low. Thus a sphere of uranium of a given mass can be less than a critical mass at a low density and more than a critical mass if it is compressed to a higher density.

Compression is achieved by means of the "implosion" mechanism. An implosion differs from an explosion in that it involves a pressure wave moving inward rather than outward. The implosion produces the compression which then causes the nuclear explosion of the bomb.

The simplest implosion device consists of a solid sphere of plutonium (or ^{235}U) surrounded by a spherical shell of chemical high explosive, usually TNT or something very similar in the form of a waxlike solid. Around the surface of the sphere are arranged a number of detonators which can all be set off at the same instant by an electrical signal or other impulse (Fig. 8–2). The chemical explosion proceeds in the form of rapidly advancing shock waves in the chemical explosive itself, waves which rapidly increase in their pressure as they converge upon the outer surface of the plutonium sphere. This is a method of making a sudden and terrifically powerful squeeze of the plutonium. Although the plutonium is a hard metal, it is actually compressed instantaneously by an appreciable amount, perhaps even to something like three-fourths of its original volume. For reasons already discussed, this means that it is suddenly well over a critical mass for the increased density (though it may have been about 0.99 of a "crit" before compression), and an enormous amount of energy is generated by the chain reaction in the brief instant before the material is blown apart.

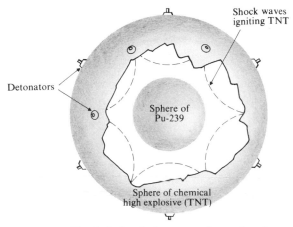

Fig. 8-2 Implosion type of atom bomb.

The simple statement that much energy is created before the bomb blows itself apart may seem obvious, particularly in view of the, perhaps unfortunate, experimental observation that A-bombs do explode. However, if one tries to understand it, it is not quite so obvious and is worth a little thought. For the amounts of energy generated, there is no material with enough tensile strength to hold the device together while the chain reaction proceeds. This can only be done by the inertia of the plutonium itself (perhaps helped by the inertia of dense material surrounding it known as a tamper). This inertia is merely a resistance to the building up of outward velocity, the factor m in the equation $F = ma$. The more rapidly the outward force is built up by rapid energy production at the center, the greater the acceleration and the shorter the time before the plutonium sphere is expanded enough to stop the chain reaction. What the chain reaction does is to release energy at a very high rate when the plutonium is near its maximum concentration. The total energy produced is the average rate of energy production multiplied by the time during which it is produced. As the rate of production gets greater, the time gets shorter because the sphere is blown apart faster.

Is anything gained by having a high rate of production? The answer is "yes": in spite of the outward acceleration the inertia does hold the material together long enough for a high rate of production to produce a large total energy. This is discussed in more detail in Appendix 9.

So the high rate of energy production of the chain reaction does produce an enormously powerful explosion.

When an atom bomb explodes, it seems from what is publicly known that only about 10 or 20 percent of the fissile nuclei in it undergo fission before the chain reaction stops. However, this is enough to release a great deal of energy. We have learned that fission releases 200 Mev per nucleus. Translated into other units, this means that, if all the nuclei in one kilogram of ^{235}U or of ^{239}Pu were to undergo fission, the energy release would be the equivalent of about 20 kilotons of TNT. This is roughly the amount of energy released in the Hiroshima bomb and in the Nagasaki bomb, as well as in the original bomb of the Nagasaki type tested as the first nuclear explosion at Alamogordo, New Mexico, in July 1945, the month before the attacks on Japan. (Actually the Hiroshima bomb's energy was less than 20 kT and that of the Nagasaki bomb slightly more than 20 kT but the difference is not very significant; it is convenient and almost customary to speak of them both as 20 kT.) Various figures in the neighborhood of 7 or 10 kilograms have been mentioned publicly as the amount of fissile material in one of these bombs. Ten kilograms with 10 percent utilization would mean 1 kg undergoing fission, the amount required to yield the 20 kT of explosive energy.

There is a fundamental limit to the size of an A-bomb, arising from the fact that the original configuration before detonation must be subcritical, so it will not explode right away. It is possible by certain tricks to increase the amount of material that constitutes a critical mass, for example, by using a hollow shell rather than a solid sphere. But there are limits to how far one can go with this. Clever engineering can also attempt to increase that 10 or 20 percent figure, but there are limits also here. Thus one cannot go on increasing the power of an A-bomb indefinitely.

The maximum that has been attained is thought to be about 250 kT (that is 250 kilotons of TNT equivalent), about a dozen times the power of the original A-bombs. The United States already had A-bombs that powerful by 1952 when H-bombs were invented. This was already a tremendously powerful military deterrent, so the political stability of the strategic balance was not as sensitive as some politicians hysterically proclaimed to the question of which side first attained the H-bomb. Actually, the U.S. and the U.S.S.R. attained it within a few months of each other in 1953. (In the last days of the Truman administration in 1952, the United States inaugurated the "thermonuclear age" by testing at Bikini in the Pacific a "thermonuclear device" which, with its bulky refrigerating apparatus, was not a true H-bomb. Unless we count this, the Russians were half a year ahead.)

THE H-BOMB

The H-bomb makes use of a sudden thermonuclear reaction, synthesizing the isotopes of hydrogen into slightly heavier nuclei. This reaction requires enormously high temperatures in order that the thermal agitation may overcome the electric repulsion between the hydrogen nuclei.

The H-bomb differs from the controlled thermonuclear reaction in that no containment is needed; it is intended to explode. As in the case of the A-bomb, the inertia of the outer parts holds it together for the reaction to proceed long enough to consume an appreciable fraction of the fuel.

The implosion of a sphere of TNT, with no other material at the center, could be used to produce a very high temperature at its center. However, this falls far short of the temperature needed to ignite a mixture of the hydrogen isotopes, to start a thermonuclear reaction. Effort has been devoted to trying to start such an explosive reaction directly from chemical energy or concentrations of electric energy without success.

However, a much greater concentration of energy than available by these means is provided by the explosion of an atomic bomb. Thus an H-bomb uses an A-bomb as its trigger or detonator to provide the high temperature needed. Aside from this, very little can be said about the geometrical design of an H-bomb. The external appearance is known and is shown in Fig. 8–3. The detailed internal plans or even sketched plans are highly secret. This is an aspect of secrecy against which there should be no political complaint. These secrets would be useful only to someone wanting to make an H-bomb, and in order to avoid the proliferation of nuclear weapons the fewer of these people there are and the harder it is for them the better.

But more is known about its nuclear physics and chemistry. The hydrogen needed (deuterium or tritium or a mixture of them) must be in a concentrated form, so refrigeration to very low temperatures can

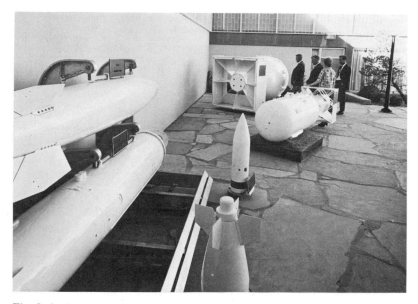

Fig. 8–3 Casing of atomic and hydrogen bombs on display at the Los Alamos Scientific Laboratory. In the background is a casing similar to the one dropped at Nagasaki (known as "Fat Man"). Its shape suggests the spherical implosion mechanism. In front of it is the type used at Hiroshima (called "Little Boy"), a gun assembly type. Each had a yield of roughly 20 kilotons. In the foreground are the more compact designs that evolved with further development. In the middle is an atomic artillery shell, the smallest of these weapons. The sleek model on the top left is a thermonuclear weapon casing, with a yield in the megaton range, several hundred times as powerful as its bulkier predecessors. The casing on the bottom is of a fission weapon.

be used to reduce it to a liquid. This would be much too awkward for a practical bomb. It was indeed done in the "thermonuclear device" tested at Bikini in 1952, an event which had resounding political repercussions because a highly respected scientist and advisor, former wartime chief of the Los Alamos Laboratory, Robert Oppenheimer, among others, considered it technically unsound as an approach to an H-bomb and was accused of political motivations for his technical advice. At about that time an essential ingredient of H-bombs was discovered both in the United States (by Edward Teller and Stanislas Ulam) and in the U.S.S.R. This was the realization that hydrogen occurs in a marvellously useful solid form in lithium hydride, LiH, and furthermore that lithium can add to the bang. Thus the chemical binding holds the hydrogen isotopes in a compact form without any cumbersome refrigeration. If the lithium in this substance were present just as a useless mixture absorbing neutrons, this could prevent the reaction. However, the lithium nuclei serve a useful purpose at the same time as the lithium atoms hold the hydrogen atoms in the solid.

A very important thermonuclear reaction between the hydrogen nuclei is

$$D + T \rightarrow \alpha + n + 17.4 \text{ Mev.}$$

(Deuterium and tritium combine to give alpha particles, neutrons, and energy. This is listed as reaction 3 in Table 3–3.) This reaction has a larger cross section than the others and starts as a thermonuclear reaction at a somewhat lower temperature than the others, still about 10^8 degrees C. At least a small amount of tritium is needed, mixed in the deuterium, to get such a reaction started, and this reaction may then further increase the temperature to make the other reactions, especially D + D, take place. Indeed, a mixture of deuterium with a little tritium, kept liquid at low temperature, was the basis of the first large thermonuclear explosive experiment. But tritium is radioactive with a half life of 12 years and would be a troublesome permanent ingredient of a bomb, making unwanted heat and the necessity of frequent replenishment.

The use of lithium deuteride, and in particular lithium deuteride with the isotope lithium-6, ^{6}LiD, solved two crucial problems at once. As a chemical molecule it holds the deuterium in solid form without the need for refrigeration, and at the instant of detonation the ^{6}Li present provides the tritium needed for the D + T reaction through the reaction

$$^6\text{Li} + n \rightarrow \alpha + T + 5 \text{ Mev.}$$

It is because this reaction has a larger cross section than the corresponding ^{7}Li reaction that the separated isotope ^{6}Li is used in the ^{6}LiD. So here suddenly we have the production of more tritium to increase the intensity of the thermonuclear reaction. Since lithium is cheap and tritium is expensive, this has economic as well as technical advantages.

We may then think of an H-bomb as consisting mainly of an A-bomb surrounded by, or close to, a mass of ^{6}LiD. When the A-bomb trigger explodes, and instantaneously before it has a chance to blow the H-bomb apart, it emits a very intense burst of neutrons that bombard the ^{6}Li to make the tritium needed, mixed with the deuterium already there, so that the heat produced by the A-bomb can detonate the D + T reaction. This in turn gives off more heat to speed up this and other reactions and at the same time makes more neutrons to hit ^{6}Li and make more tritium, speeding up the D + T reaction.

All this interaction between the reactions conspires to make the explosion proceed very suddenly and to attain a high energy release, using up a considerable part of the materials present before the explosive expansion proceeds far enough to stop the reaction. Diagrammatically, we may envisage the interaction of the reactions as follows:

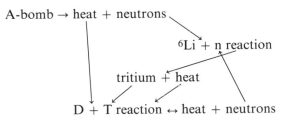

The heat from the ^{6}Li + n reaction is probably much less important than that from the A-bomb in getting the reaction started. Indeed, once these reactions attain a very high temperature, other reactions come into play to contribute to the total yield of energy, particularly the reaction

$$^6\text{Li} + \text{D} \rightarrow 2\alpha + 22 \text{ Mev.}$$

This has a large energy release but requires an especially high temperature because of the electric repulsion of the lithium with its $Z = 3$.

A large part of the energy of the explosion comes from that 22 Mev and from the 17.4 Mev in the case of the D + T reaction. These are each only about a tenth as much as the 200 Mev released in the fission reaction, but hydrogen and lithium are so much lighter than uranium that on a weight basis the energy yield per kilogram is greater in an H-bomb. A more important reason for the greater yield per kilogram in an H-bomb is probably that the reaction consumes a larger fraction of the fuel before being stopped by the expansion. This could arise both from the fact that reactions dependent on the highest-energy atoms in thermal agitation become rapidly more intense as the temperature rises and from the way the production of new fuel (tritium) by the reaction itself suddenly takes the material far beyond the critically explosive condition.

As in the case of electric power, so also in the case of bombs, there are just two kinds of sources of great amounts of energy, fission and fusion. Figure 3–14 shows that the internal energy of a nucleus, per nucleon, is lower for intermediate-mass nuclei than for either very heavy nuclei or very light nuclei. As we have seen, fission exploits this relation from the heavy-nucleus end. Fusion exploits it from the light-nucleus end. As was said earlier, for the sake of political judgments, it is important to realize that there are not any more tricks to be pulled out of this bag.

"CLEAN" AND "DIRTY" BOMBS

In the international political developments of the mid-1950's, the hope for the development of a "clean" bomb played a considerable role as an argument against stopping testing by international agreement.

The production of energy by fusion creates much less radioactivity than does the creation of energy by fission. Any H-bomb has a fission trigger and so releases at least the fission products from a fairly "small" A-bomb. However, if it is a large H-bomb, by far the greater part of its energy comes from fusion, and the release of radioactive fission products per kiloton of yield is relatively small. It is so tremendously more powerful than an A-bomb that the radioactivity per kiloton of yield is small, and it may be called "relatively" clean.

There is another way to reduce radioactivity. Even though the A-bomb used as a trigger must have a near-critical mass, it might be possible to design the trigger mechanism so cleverly that less than the maximum power of the A-bomb is needed. The A-bomb could then be designed to have less than normal efficiency and less than normal yield of radioactivity from fission products. Thus the fission yield of a "clean" bomb may be somewhat less, but not very much less, than that of a normal A-bomb. In addition to its fission yield, there is some unburned tritium to keep it from being really clean.

However, from the tests that have been carried out, it is surmised that many if not most of the H-bombs in weapons stockpiles are "dirty" bombs. This is an economical type of bomb made by wrapping a blanket of ^{238}U around an H-bomb. This serves a double purpose, first as a tamper providing the inertia that holds the bomb together to react a little longer, and second as a tremendous source of fission energy as a result of the fast neutrons produced in the thermonuclear reactions:

$$D + T \rightarrow \alpha + n + 17.6 \text{ Mev}$$
$$D + D \rightarrow {}^3He + n + 3 \text{ Mev}$$

$$^{238}U + \text{fast neutron} \rightarrow \text{fission}$$

This isotope, while not as easily fissile as ^{235}U, does have a fast-neutron cross section sufficient for this purpose. Uranium-238 is cheap because it is a by-product after the valuable ^{235}U is separated from it in the gaseous diffusion plant. As much of it may be added as the delivery vehicle can carry; there is no critical-mass limit. Thus even in an H-bomb of this dirty variety, most of the energy may come from fission with its consequent enormous yield of radioactive fission products. This is known as a fission-fusion-fission bomb.

DESTRUCTIVE POWER OF NUCLEAR WEAPONS

Nuclear weapons of course get their military effectiveness from their enormous destructive power.[1,2] This comes in three categories: blast, direct radiation, and radioactive fallout.

At Hiroshima and Nagasaki the A-bombs, with yields approximately 20 kT, were each exploded about half a mile in the air over the center of the city. The serious blast damage, knocking down brick buildings, extended to almost one mile. This was also about the radius at which radiation burns were very serious. Some victims died of radiation burns and attendant radiation sickness, others with smaller portions of their skin exposed recovered.

The radius at which blast is effective increases with the yield Y of the bomb but not very rapidly. The relationship is

$$Y = \text{constant} \times r^3.$$

This means that the yield has to be multiplied by eight to increase the radius by a factor of 2, or to increase the area of severe blast destruction by a factor of 4. Thus, from the point of view of increasing the area of blast destruction it does not pay to go to extremely powerful weapons. For example, four 1-megaton weapons would have the same area of severe blast damage as one 8-megaton weapon. By way of explanation, we can understand that the 8-megaton weapon would "overkill" the central region, pulverizing it more energetically than does the less powerful weapon. This is one reason, though not the main reason, why military men are now interested in having several smaller warheads rather than one big warhead on a single intercontinental missile.

About half of the energy of an air-burst H-bomb goes into blast, about a third into thermal radiation (heat), and the rest into prompt and delayed nuclear radiations (prompt gamma rays and neutrons and the delayed radioactivity of fission products). The prompt radiations cause biological effect to those exposed in the form of skin burns or radiation sickness, but these do not extend much beyond the radius of severe blast damage so for the most part affect victims who would be killed in any case. There were survivors of the blast in the bursts over Japan in whom these effects lingered, but they would probably be proportionately fewer with larger bursts.

The radiation from big bombs not only causes somatic damage but also starts fires. In clear weather this is the most destructive aspect of big H-bombs if they are ever used against cities. Experience with incendiary raids on German cities in World War II showed that when many fires are started in one locality the result is a firestorm. It is like putting more logs on the fire in the fireplace, each log helps the others burn. In the case of a large city with many fires started, the rising heat causes a wind of hurricane velocity toward the center of the city, fanning the fires and causing them to spread from one tree or building to the next. This can engulf the entire city in flames, burning practically all the people, or suffocating them if they have taken refuge in sufficiently deep shelters to avoid burning (unless oxygen is provided).

The distance at which the radiation starts fires in clear weather, r_{fire}, varies in a somewhat different way with the explosive yield of a bomb burst high in the air, as indicated in Table 8–1.

The yield required for destruction by firestorm is roughly proportional to the square of the radius, r^2, rather than r^3, and thus the area per megaton is roughly constant for large bursts in the megaton range.

The 20-mile figure in the table, for example, is intermediate between the 25-mile radius at which light trash, such as exists even in well-kept suburbs, is ignited and 13 miles at which the side of a frame house bursts into flames.   These are tremendous areas of destruction.   An H-bomb of the fairly modest yield of 1 MT can destroy by firestorm practically everything and everyone within a radius of about 7 miles,

TABLE 8–1
Extent of bomb damage

Explosive yield, Y	20 kT	1 MT	10 MT
Severe blast damage, r_{blast}	1 mile	4 miles	8 miles
Area per MT, in mile^2/MT	$\dfrac{\pi}{1/50} = 150$	50	20
Firestorm radius, r_{fire}		7 miles	20 miles
Area per MT, in mile^2/MT		150	130
Ratio of areas, $(r_{fire}/r_{blast})^2$		3	$6\frac{1}{2}$

or in a circle 14 miles in diameter having an area $\pi r^2 = 150$ square miles (about three times the area of the circle destroyed by blast).   Thus a single 1-MT bomb could destroy quite a large metropolitan area and kill almost everybody in it, though it might take a 10-MT bomb, or about four 1-MT bombs, to take out New York or Chicago with their principal suburbs.   Military estimates of immediate casualties usually count mainly on the blast (as would apply on a wet foggy day) and would conclude that about three times as many bombs as this would be needed.

These figures of the enormous destruction by a few bombs are to be read against an awareness of the approximate size of the weapons stockpiles of the two "nuclear giants."   For missile delivery alone, the United States has over 1000 land-based intercontinental missiles with about 3-MT thermonuclear warheads and about 650 submarine-launched missiles with warheads somewhat over 1 MT.   These figures are before installation of multiple warheads.   In addition to this, alerted bombers are equipped to deliver 2100 bombs which constitute by far the largest part of the deliverable megatonnage.   The U.S.S.R. has somewhat smaller numbers of somewhat more powerful missile warheads (corresponding to the greater urban sprawl of their potential United States targets).

FALLOUT

The lethal effect of the fallout in a hypothetical nuclear war depends heavily on whether the bombs are burst high in the air, as in attack on cities for the sake of maximizing blast damage (preventing one building from shielding another), or whether it is a ground burst as in an attack on missile sites.[6] In the first few seconds the concentrated energy of a nuclear explosion expands as a very hot fireball, glowing brighter than the sun (Fig. 8–4). This vaporizes whatever it comes in contact with, including the surface of the ground if it is low enough. The fireball has a radius depending on the yield:

1 kT	200 feet
1 MT	2000 feet
10 MT	4000 feet (about 3/4 mile)

Air bursts designed to maximize blast damage would be at an altitude varying from about half a mile for 20-kT A-bombs to two miles or more for H-bombs in the megaton range. Such air bursts are thus plenty high enough that the fireball does not touch the ground, and the tremendous updraft carries the fission products and the rest of the vaporized bomb high into the stratosphere, fanning out at the top into the famous mushroom cloud. Part of this radioactive bomb debris falls as local fallout and is distributed downwind in a cigar-shaped pattern which after a few hours may be several hundred miles long and a few tens of miles wide, decreasing in intensity as it gets larger and older. Of the fission products that stay aloft longer, much is brought down by rain, some of it quite soon and some years later.

A ground burst, such as would occur in an attack on missile sites, makes much more severe local fallout. The fireball spreads out over the ground. The vaporized earth condenses into small particles as it cools and brings down most of the radioactivity as very intense local fallout. Only about 20 percent is carried to the stratosphere. Thus an attacker can choose to emphasize damage and death due either to blast and firestorm or to radioactivity.

About 1959 there was extensive discussion and even promotion of a national fallout shelter program to mitigate this aspect of the destruc-

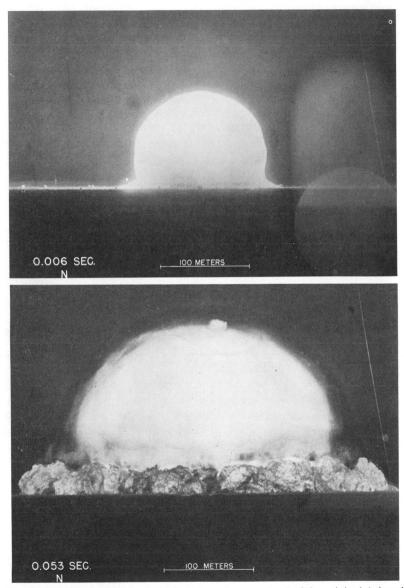

Fig. 8–4(a) Growth of the fireball and mushroom cloud of the original A-bomb test in New Mexico.

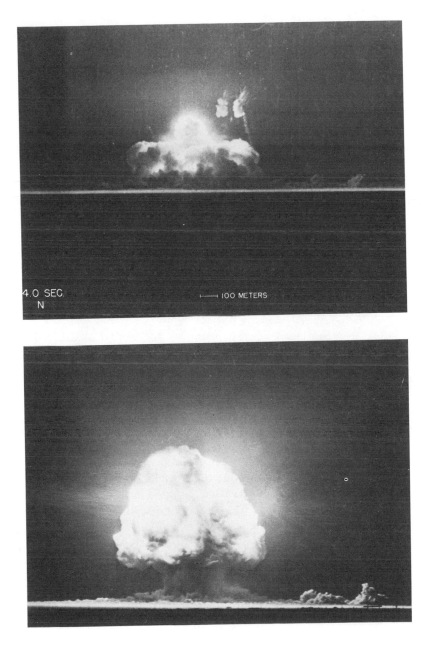

Fig. 8–4(b)   The mushroom cloud of a later test.

tion in a possible nuclear war, partly with the hope of making the war less likely by making the country appear to be a less vulnerable target. In Congress the Joint Committee on Atomic Energy held hearings about the effects of a postulated attack on the United States, carefully described as involving 263 warheads for a total of 1446 MT (half fission, half fusion) in a definite pattern about equally divided between cities and military targets. There was in the subsequent discussion a tendency to consider these oddly precise figures as describing a typical rather than a rather small prospective nuclear war.   About a quarter of the dwellings in the United States were considered destroyed by blast (firestorm neglected) and about that many subjected to radiation activity of more than 100 roentgens per hour from fallout at the end of the first hour.   The radiation level from fallout would still be above 1 rad/hr in 15 percent of the land area after 2 days. It would be above 0.1 rad/hr in about half of the country after 2 days and in 15 percent of the country after 2 weeks.   Thus even in this 15 percent of the land area, which tends to be the most populous part, people could come out of shelters after two weeks for short periods and try to wash away some of the radioactivity—if they could fetch water from the brook when they find it does not flow from faucets because of lack of electricity to pump water.   From this hypothetical moderate-sized nuclear war, strontium-90 levels throughout the northern hemisphere were estimated as about 1 curie per acre, about ten times that experienced from the weapons testing of the late fifties.[17]   (Each megaton of fission energy creates 10^5 curies of strontium-90.)

It is possible that a nuclear attack might be made in such a way as to employ fallout as the principal mode of destruction. For example, if cities could be defended by an effective local defense system, ground bursts of large dirty weapons to windward of cities could be used to attack their populations.   Ground bursts attacking the numerous missile sites in the remote regions of Montana and North Dakota would have some of this effect on the cities to the east of them, though the distances are too great for maximum effectiveness. A mode of attack not requiring advanced missile technology would be to float a few very large bombs by means of ordinary vessels a few miles or even a few hundred miles off our west coast. With sufficiently large and especially prepared bombs this could be an attack on the entire country, but in this case would probably have radioactive repercussions on the attacking country as well, unless it were in the southern hemisphere.

In this connection the hypothetical cobalt bomb is of interest. Cobalt is a metal whose nucleus ^{59}Co absorbs neutrons very freely to make radioactive ^{60}Co with a five-year half life. We have learned of its important uses in medicine and industry, but it is also an exceptionally effective material for making a dirty bomb. It can be wrapped around a large hydrogen bomb in almost unlimited amounts to absorb the superfluous neutrons and produce fallout enormously more potent than that from an ordinary atomic bomb.

Serious discussions of the psychology of national leaders faced with the threat of nuclear war have included consideration of the hypothetical concept of a "doomsday machine."   This is a sort of reductio ad absurdum of the nuclear arms race: the hypothetical instrument of destruction so powerful that a national leader by pushing a button could destroy all life on earth. Cobalt bombs have been mentioned as possibly the main ingredient of a doomsday machine. A popular novel attempting to bring to the public an impression of the awful relentlessness of nuclear war, *On the Beach,** used a cobalt bomb as its starting point but substituted a mild euthanasia of gentle death for the full horror of nuclear war. Technically, it also ignored the fact that the winds of the two hemispheres mix almost not at all.

Between blast, firestorm and fallout, the destructive power of nuclear weapons is so vast as to be almost beyond comprehension. It is "unthinkable" and yet something that must be thought about if policies are to be guided wisely to minimize the likelihood that the possible devastation will come to pass. Back in 1947, when nuclear weapons meant 20-kT A-bombs, former Secretary of War Henry Stimson wrote, "The bombs dropped on Hiroshima and Nagasaki ended the war. They also make it wholly clear that we must never have another war. This is the lesson men and leaders everywhere must learn. . . . There is no other choice." (See Appendix 16.) Since then the power of the largest weapon has been multiplied a thousandfold, its destructive effect a hundredfold. The explosive power of the world's stockpiles of nuclear materials has multiplied many tens of thousands of times. The 1947 stockpile is not even visible on the scale of the graph of Fig. 8–5 that shows the dramatic increase in the late fifties and early sixties when the isotope plants built as a reaction to the invention of the H-bomb came into full production. Responsible leaders are deeply convinced there must never be a nuclear war. (They have already been known to risk nuclear war but with the intention of avoiding it, as in the Cuban missile crisis of 1961.)

DETERRENCE

Yet nations are piling up ever greater means of nuclear destruction in order not to have to use them. Each of the two nuclear giants, the United States and the U.S.S.R., is doing its utmost to keep up its end of the balance of terror. For the short term this seems to each the safest thing to do. Decisions are commonly made for the short term.

---

* Neville Shute, *On the Beach*, New York, Morrow, 1957.

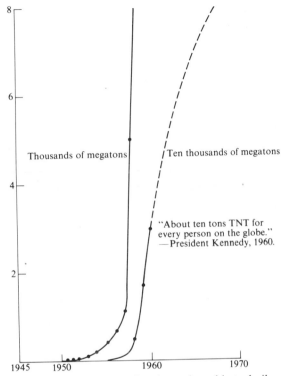

Fig. 8–5 Growth of explosive power of estimated world stockpile of nuclear weapons materials.

For the long term it seems to mean a terrible risk.

Once two great nations had nuclear weapons, each apparently based its policy on the assumption that the other might attack if it had a chance to do so without nuclear retribution. This is the assumption that underlies what is considered responsible military planning. As former Secretary of Defense Robert McNamara put it in 1967, "When calculating the force we require, we must be 'conservative' in all our estimates of both a potential aggressor's capabilities and his intentions. Security depends upon taking a 'worst plausible case'—and having the capability to cope with that eventuality." (Appendix 18)

Since there is no adequate military defense against a massive nuclear attack, the only military way to prevent the attack is to have the capability of retaliating with such devastation that the adversary cannot consider it "acceptable." One must not only have this capability but must convince the adversary that the retaliation is inevitable in case of attack. This situation, in which each nation by threat of retaliating terror denies the other the option to attack with impunity, is "deterrence." This is the "balance of terror."

In principle, between nations with rational leaders, deterrence should successfully prevent attack as long as neither side has such great superiority in nuclear weapons that it could knock out almost all of the adversary's retaliatory force and thus reduce the retaliation to an "acceptable" level. The destructive power of just one H-bomb is so terrific—there is so much prospective horror and suffering and disruption of economic life behind those cold statistics of the number of square miles devastated—that one might think offhand that the destruction of one great city by one H-bomb would constitute unacceptable damage in the mind of any rational national leader. Yet Japan in 1944–45 absorbed an amount of terror comparable to that in the fire raids that preceded the A-bombs and was still fighting. The U.S.S.R. suffered 22 million casualties in the course of resisting a German invasion. In neither case was such punishment anticipated in making decisions that led to the war, so those events were not really relevant to the kind of decision that would have to be made to make a first nuclear strike. Nevertheless, the law of the "worst plausible case" leads to the assumption that devastation by many H-bombs would be required to deter.

FIRST-STRIKE AND SECOND-STRIKE CAPABILITY

The two nuclear giants face each other, each with large numbers of land-based long-range missiles, submarine-launched missiles of somewhat shorter range, and manned bombers carrying nuclear weapons. Each is presumably on constant lookout for a surprise attack by the other. Under the present circumstances, the land-based long-range missiles (ICBMS) would presumably be the principal instrument of surprise attack, reaching to the continental interior with sufficient accuracy to destroy ICBMs in their underground protective silos. Each side has radar with which to observe and track incoming missiles. It takes twenty minutes or half an hour for an ICBM to travel to its target 6000 or 8000 miles away. The radar might observe it when it is still 10 or 15 minutes away. In case of a hypothetical attack, the attacked country would have to decide in this short time whether to send off its ICBMs in retaliation before the attack arrives to destroy the ICBMs in the ground. If the policy is to be this trigger-happy, a false interpretation of a radar signal could initiate an accidental nuclear war. This is one possible way that deterrence can be unstable.

The nuclear giants recognize that it would be better to have confidence in being able to retaliate even after receiving the attack, so that the decision need not be made quickly on flimsy evidence. The diversity of forces and the fact that the submarine-launched missiles would presumably be available for retaliation even if all the land-based missiles and bombers were wiped out should help make the trigger-happy response seem unnecessary.

A nation is said to have second-strike capability if it can absorb the entire force of its adversary's surprise first strike and still have the means for launching a second strike that would inflict unacceptable damage in retaliation. That is, it has the capability to wait and make a second strike deliberately. In this situation the adversary does not have first-strike capability; he cannot strike first and get away with it. He can make a first strike but only as an act of near-suicide. Deterrence is considered relatively safely established as long as both sides have second-strike capability and neither has first-strike capability. Throughout the two decades since the early fifties this has been the case, as force levels on both sides have increased drastically.

OVERKILL

In 1951, just before the invention of H-bombs, a hundred A-bombs would have been considered quite enough to deter. The U.S.S.R. had less than that and questionable delivery capability; yet it was already too late for the United States to consider seriously risking a preventive first strike to avoid the arms race. A decade later both sides had destructive power hundreds of times greater in H-bombs deliverable by missiles as well as long-range bombers. (The United States had about a hundred H-bomb-carrying missiles and the U.S.S.R. about a quarter that many.) Surely this must have been more than adequate to deter. Yet nuclear arsenals have gone on increasing in power and sophistication. [3,4]

Military establishments, traditionally charged with being capable of fighting and winning wars, understandably react to the new charge of being able to deter by placing what seem like high numbers on the needs for deterrence. Secretary of Defense Robert S. McNamara had eight years of contention with his Joint Chiefs of Staff over this point, trying with some success in some categories to moderate their demands. In 1967 the Department of Defense gave estimates of damage to the U.S.S.R. in terms of delivered megatons shown in Table 8–2.

The total population fatalities are reckoned only as immediate deaths (or within 24 hours) and are thus a gross underestimate of the effects of a nuclear attack. It is, however, very difficult to put numbers on the expected later deaths due to radiation sickness plus general disease and starvation in the unprecedented utter chaos following the

TABLE 8–2

Estimates of capability to destroy Soviet population and industry

(Assumed 1972 total population of 247 million, urban population of 116 million)

1 MT equiv. delivered warheads	Total population fatalities		Industrial capacity destroyed (percent)
	Millions	Percent	
100	37	15	59
200	52	21	72
400	74	30	76
800	96	39	77
1200	109	44	77
1600	116	47	77

sudden widespread disruption of normal facilities. Secretary McNamara underlined the 400 MT figure and said "beyond 400 one-megaton equivalents optimally delivered, further increments would not meaningfully change the amount of damage inflicted because we would be bringing smaller and smaller cities under attack."

He did not say that anything like this number is needed for deterrence, but rather that this is a number beyond which one would not make more meaningful damage. One would reach smaller cities and towns, but all major cities and most factories would have been completely destroyed. Yet we have gone far beyond this number in our arsenal. We have enough missiles (before multiple warheads and not counting those in Europe) to deliver those 400 one-megaton-equivalent bombs to destroy all that is meaningful in the U.S.S.R., and then to repeat the bombardment three times, and then to do it four or five times over again with bombers. This is the kind of capability that goes by the name of overkill.

The United States delivery vehicles on which this capability is based were as follows (as of 1970):

1. 1054 ICBMs, of which 54 were Titans with 5-MT warheads, and about 1000 Minutemen with 1- to 2-MT warheads.

2. 656 Polaris missiles in 41 submarines, each missile with almost 1-MT equivalent.

3. 550 long-range bombers, each (with a few exceptions) carrying four 5-MT bombs.

4. 2000 tactical aircraft (not intended for strategic attack but capable of reaching the U.S.S.R.), each with a 5-MT bomb.

The Soviets have fewer delivery vehicles to send against us, but in some cases with considerably more powerful warheads. While perhaps they are not able to kill us quite so many times over, still they have overkill capability. These figures are given in more detail, with a graphic history of their growth and projections to 1975, in Appendix 17.

In contemplating how we ever came to take the risk and go to the expense of piling up such tremendous means of destruction, we must understand first that the United States has led the way all the way, so we must look to our own psychology and politics and economics for the answer. Not that the U.S.S.R. has been innocent of bellicose ways, but it is our reaction to them that is relevant for the overkill buildup. In nuclear weaponry they have struggled to try to keep up with us within a factor of two or three or four, thereby denying us first-strike capability.

Deterrence has remained apparently stable with this great disparity of destructive power between the two sides. This illustrates how great is the stability of the balance of terror between rational adversaries, and that it is not necessary for one side to keep ahead to deter attack by the other. While we of course impute innocent intentions to ourselves, it has been admitted militarily all along that even with our impressive lead we did not have first-strike capability, regardless of our motives.

The competitive urge in American psychology, the human desire to win any race, filled us with an irrational compulsion to stay comfortably ahead. But in this game we never knew what the score was, and this ignorance gave full sway to political pressures to keep pushing ahead every time there was a decision to be made and to grant every Department of Defense request. Throughout the late fifties and early sixties when the buildup was rapidly gaining momentum, secrecy about the size of our nuclear arsenal played an important influence on appropriations. This secrecy seemed to have no other rationale, if our aim was really deterrence. The aim in deterrence should be to convince the adversary that one has sufficient retaliatory forces and the will to use them if necessary. Keeping their number secret when we had actual superiority would seem to have the opposite effect.

The purpose of the secrecy may have been to preserve the possibility of bluffing the other side into thinking we had more than we did, in case they should make a breakthrough to some new kind of super-weapon that would leave us behind in the race. But the practical effect of the secrecy seems to have been to permit the Soviets to learn the approximate size of our stockpile through espionage (in which they might have more confidence than in press releases) while Congress and the American people remained in the dark and gullible at appropriations time. Not only the number of weapons but even the amount of fissionable material on hand was a carefully guarded secret. although the latter could be estimated from such data as the power input into the thermal diffusion plants.[3,4]

Another important ingredient of the momentum of arms procurement is the fact that arms production makes industrial contracts and jobs, and industrialists and job-holders elect political "leaders" who have to be "followers" to be elected.[4,16] Not until the remarkable 50–50 vote in the Senate on August 6, 1969, when a particularly dubious antiballistic-missile system was at stake, was there any serious challenge in Congress to a military appropriations bill. It was widely recognized that Congress had become a rubber stamp for the desires of the Department of Defense.

Another aspect of the arms momentum arises from dissension within the Department of Defense. There is rivalry between Army, Navy, and Air Force that seems to require giving matching plums to each. There has been disagreement over doctrine, including a reluctance to accept mere deterrence as the Department's nuclear mission. In testifying before a Congressional subcommittee in 1968, the Chairman of the Joint Chiefs of Staff, General Wheeler, said:

> Our national security objective is "to preserve the United States as a free and independent nation, safeguarding its fundamental values, and to preserve its world power." From this we produce our basic military objective, to deter agression at any level and, should deterrence fail, to terminate hostilities, in concert with our allies, under conditions of relative advantage while limiting damage to the United States and minimizing damage to the United States and allied interests.

The idea that we should in some sense be prepared to "win" a nuclear war with some vestiges of survival and recognizability as a nation can be related to a bizarre extension of a policy of deterrence thus: An adversary not deterred by the prospect of being practically annihilated in retaliation might still consider it less worthwhile to attack us if there were some prospect that we might survive better than he would as a recognizable nation. The buildup of our strategic striking forces to the point of multiple overkill may be more closely related to a military desire to be able to annihilate an enemy thoroughly and dependably if the occasion should arise than it is to simple deterrence.

TACTICAL NUCLEAR WEAPONS

Strategic bombing apparently got its name from the World War II "strategy" of bombing an enemy homeland to reduce his ability to continue hostilities. Strategic nuclear weapons are those long-range weapons intended primarily for attack on an adversary's land area. Tactical nuclear weapons, by contrast, are those intended primarily for use in the "battlefield," to interdict the movement of troops and the like. The answer of the United States and its NATO allies to the preponderance of Russian land armies, with their many tanks, was to declare nuclear weapons of something like the power of the Nagasaki bomb, 20 kT, and smaller to be tactical nuclear weapons. As developed, they have become quite compact. Figure 8–3 shows how compact all the new nuclear weapons are compared to the World War II originals. The smallest is fired as an 8-inch artillery shell, but even this small shell is said to have a power up to 20 kT. For some situations a tactical nuclear weapon may have a power of 1 kT or less.

The advantage of such weapons is that they can be considered to deter a conventional armed invasion with a minimum of expensive troop deployment. This is the reason United States tactical nuclear weapons are deployed in some European countries (notably not France, who has chosen to go her own way in nuclear weaponry). United States law requires that the weapons remain under United States control; they must not be given over to the control of allies except in the emergency of actual attack. Yet our NATO allies in many cases control the means of delivery. This requires that United States control be reduced to a rather perfunctory arrangement sometimes called "the key in the Colonel's pocket." For example, West German aircraft crews are prepared to take off at a moment's notice and drop American tactical nuclear bombs, but the bombs are locked up and would become available to them only when unlocked by an American officer who remains on the spot. It seems an insecure arrangement. The officer could easily be overwhelmed, but only as an affront to the United States.

The main disadvantage of such weapons is that if they are ever used it would be natural for the enemy to use somewhat larger weapons in retaliation, and there is no natural limit to the escalation right up to full-scale nuclear war. So long as only conventional weapons are used, there is a clear line of demarcation. In the nuclear age it seems too dangerous to fight any wars, largely for fear of this escalation. This is probably what former Secretary of War Stimson meant when he said that the Hiroshima and Nagasaki bombs "make it wholly clear that we must never have another war." (See Appendix 16.) Unfortunately, the world has not so concluded, but the nuclear powers have at least taken care not to come into direct confrontation themselves, participating partly vicariously in wars that have not become world wars. In these it is enormously important that tactical nuclear weapons should never be regarded as just another weapon in the arsenal. As long as the clear line between conventional and nuclear weapons is not broached, there remains hope that all-out nuclear war can be avoided. Once that line of demarcation is passed, the sky would be the limit.

A secondary disadvantage of tactical nuclear weapons is that the region "defended" by them would be apt to become "scorched earth," even if there were no escalation to still more powerful weapons.

A "low-yield" nuclear weapon cannot be made with arbitrarily little fissionable material, because there is a critical mass below which the material will not sustain a chain reaction and explode. The critical mass can be reduced by compression to somewhat less than the mass needed at normal density, but something like five kilograms of ^{235}U or ^{239}Pu has to be used no matter how small the desired yield may be. A 1-kiloton bomb uses just about as much fissile material as does a 20-kT bomb, but uses it less efficiently. In this sense a low-yield tactical weapon is purposefully a "fizzle." The term low-yield applied to a nuclear weapon is misleading. A 1-kiloton nuclear artillery shell would have 20,000 times the power of a conventional shell of similar size carrying 100 lb of TNT.

While the number of tactical nuclear weapons the United States has deployed around the world is not divulged, it must be enormous, presumably 7000 in Europe alone. In 1964, as our stockpile was rapidly increasing, Senator Pastore of the J.C.A.E. said, "Today we count our nuclear weapons in the tens of thousands." At that time the number of deliverable strategic warheads, mostly bombs to be delivered by bombers, was in the neighborhood of two thousand, which gives a hint of how many tactical weapons we already had even then. It is doubtful whether anyone could account for the location of all of them, and it is a tribute to the reliability of the safety catches that there have been no explosive accidents.

ACCIDENTAL WAR AND CATALYTIC WAR

One serious question about the stability of deterrence is the possibility that an accidental launch of an ICBM might touch off a full-scale nuclear exchange. As the size of the deterrent force increases, this risk would seem to grow larger. When thousands of missiles in five countries must be ready for instant firing on a moment's notice year in and year out, with precautions that for this reason must not be very restrictive, it seems not beyond the realm of possibility that one missile sometime will take off aimed for some important city. One of the purposes of the "red telephone" between the White House and the Kremlin is to be able, in such an emergency, to call and say "Oops, sorry" in the twenty minutes before the errant missile hits and thus hope not to touch off a retaliatory strike.

In the United States only the President has the authority to initiate the launching of nuclear weapons. Even while traveling he can always instantly communicate with the nuclear command headquarters that can send coded orders to the various launch sites. At the launch sites of strategic weapons, both on land and at sea, the precaution is taken that no one man can alone fire a nuclear weapon. In the control room at an ICBM silo, for example, two men chosen for reliability must insert separate keys or push buttons on opposite sides of the room. Thus one man going mad could not start a nuclear war; it requires a small conspiracy.

The precaution of requiring two men to fire a nuclear missile is particularly important and perhaps inadequate in view of the unfortunate circumstance that the armed services are unable to control use of hallucinating drugs even among presumably responsible personnel. For example, when a Cuban defector brought a MIG plane through the radar defenses and landed at a base in Florida in 1969, there was an investigation and 35 men assigned to the Nike-Hercules missile crews were arrested for using, among other things, LSD. Admiral W. P. Mack, in charge of the Pentagon's drug abuse program, explained that "only ten had responsible positions—and there was only one per battery. In other words, no battery had more than one case in it."[9]

It is disturbing to think of the fantastic degree of perfection that is required when two or more countries each has many batteries of nuclear missiles, each of which must be ready for instant firing by men kept perpetually on the alert, year after year, while avoiding a dangerous degree of boredom, without even one accident. This is one reason a low-level deterrent, involving many fewer missiles, would be a safer deterrent.

It is to be hoped that all nations possessing nuclear weapons will exercise at least such precautions as requiring a minimum of two men to fire one, and will make sure they have a high degree of technical perfection in the controls. One of the dangers of proliferation is that not all nations will take such precautions.

If an ICBM were ever accidentally launched, it would proceed inexorably on its way to its prearranged target. One possible precaution against this event would be to equip the ICBMs of all the nuclear powers with in-flight destruction mechanisms to be fired on special signal. This is not done. The reason that it is not done is through fear that the adversary might learn how to duplicate the secret signal and thus be able to nullify incoming missiles. If we should suddenly become aware that one of our missiles had accidentally taken off toward a Soviet city, there would be twenty minutes or so when we would know it was on its way and could do nothing to stop the calamity of its arrival. One legitimate argument for a very "light" antiballistic missile system in all nuclear countries, able to stop one or two incoming missiles but no more, is that it might save the day in such an emergency. The trouble with deploying a light ABM system for this purpose is that military installations tend inevitably to grow. A "heavy" system would introduce performance uncertainties that would destabilize the deterrent balance. This is discussed in the next section.

Another way an all-out nuclear war could be started unintentionally is in what is known as catalytic war. The name comes from chemistry, from the way a relatively few molecules of one kind can trigger or "catalyze" a reaction among a much greater number of other molecules. Suppose in a local war one side feels hard enough pressed to resort to the use of tactical nuclear weapons. The other side would probably respond with larger nuclear weapons, perhaps still tactical. The succession of responses, perhaps bringing more powerful allies into the picture, could all too easily result in the full horror of an all-out nuclear war. This is a crucial reason for avoiding the use of tactical nuclear weapons, leaving a qualitative and easily recognizable gap between the chemical weapons that are used and the nuclear weapons that are not used by both sides in a conventional war, if there must be conventional wars.

Many analysts believe that the very existence of nuclear weapons is a crucial reason for avoiding all wars, even "brush-fire" wars, for there is the danger that one side may feel driven to extricate itself from an untenable situation by use of a nuclear weapon, whereupon the small war may grow into a nuclear war, a "catalytic" war. National policy unfortunately seems not to be guided by this view. On the contrary, there is evidence that possession of a back-up nuclear arsenal encourages a nation to undertake nonnuclear military operations that it would otherwise avoid. This use of a nuclear arsenal as a source of active national power must encourage nonnuclear nations to aspire to nuclear status.

Use of nuclear weapons in Southeast Asia has been considered at several times. President Eisenhower vetoed a proposal, supported among others by Vice President Nixon, to rescue the French with atomic bombs from what turned out to be their final defeat at Dien Bien Phu in 1954. When subsequent United States involvement in Vietnam had grown to the point, in 1961, at which policy was made in secret to involve American troops in war there, the possibility of resorting to tactical nuclear weapons in case of trouble was an explicit part of the planning, particularly on the part of the Joint Chiefs of Staff and other military authorities. These men seem to have had at least as much influence as the civil Kennedy administration in the decision to start the war, according to the revelations of the "Pentagon Papers," made public in mid-1971. For example, the commander in chief of the forces in the Pacific, Admiral Felt, is quoted as having said, "Plans were drawn on the assumption that tactical nuclear weapons would be used if required. . . ." That is, they would be used in case of North Vietnamese and Chinese invasion of South Vietnam.[16] This makes it appear that the Vietnam war could be undertaken so far from American shores partly because it was under the ultimate cover of United States nuclear superiority over China.

WEAPONS AND COUNTERMEASURES

> I was happy, I could whistle
> Until he made his anti-missile.
> I felt better when I read
> Anti-antis were ahead.
> Now I'm safe again, but can't he
> Make an anti-anti-anti?
>
> Lenore Marshall

Ever since the shield against the sword, countermeasures have been developed to blunt the effectiveness of military weapons systems. The development of antiaircraft artillery and intercepter aircraft against manned bombers are prime examples. The vulnerability and effectiveness of bombers must be considered when planning their missions. Bombs with chemical explosives were frequently as big as 10 tons and reached a maximum of 20 tons of TNT in World War II, and 15-ton bombs have been used in Southeast Asia. While formidable, their effectiveness was still sufficiently limited that from the military point of view it did not pay to deliver them if more than 5 or 10 percent of the bombers in a raid were destroyed. Thus defense was effective in the long run in World War II if it could kill 10 percent of the attack.

With the advent of the terrible threat of intercontinental missiles carrying thermonuclear warheads (that is, H-bombs fitted to missiles) it is natural that vigorous attempts should be made to develop similar countermeasures. Here, however, near-perfect interception would be needed to be effective. If as much as 10 percent or even 1 percent of the missiles of a heavy attack should get through, the result would be unprecedented destruction, even if somewhat less complete than if 100 percent of the attack missiles got through.

The direct military answer to the intercontinental ballistic missile (ICBM) would be the antiballistic missile (ABM) if it could be made to work effectively. Although Khrushchev once bragged that he could "hit a fly in the sky," ABMs cannot be made to come so close to ICBMs as to actually hit them in flight or even to destroy them with chemical explosions. The hope is to stop a nuclear warhead with a nuclear explosion, so ABMs as well as ICBMs carry nuclear warheads. One has to imagine a Hiroshima-type explosion, either high in the atmosphere or above it, trying to stop an incoming missile that, if it should get through, would cause even greater destruction than at Hiroshima. Then one must imagine many of these interceptions in quick succession, almost all at once.

ICBMs travel six or eight thousand miles, almost all of it in the vacuum of outer space. Atmospheric pressure gradually fades off with altitude in the first few tens of miles so that at an altitude of a hundred miles it is close to the perfect vacuum of interstellar space. An ICBM is called "ballistic" because it receives its acceleration and guidance in the first few hundred miles, after which it follows the approximately parabolic path of a thrown ball or artillery shell, but more perfectly because it has no air resistance. (Its path is actually a short section of an ellipse.) The ballistic path, observed by radar, can be accurately predicted for purposes of aiming an ABM to intercept it.

When ABMs were being developed as a countermeasure, it was natural to think of counter-countermeasures. These take the form of trying to "spoof" the ABM system, confusing the radar with many false signals so that they will not discriminate the true ICBM. A simple type of such counter-countermeasure is "chaff." This consists of arranging that the ICBM, shortly after it reaches free space above the atmosphere, will spew out large numbers of very light objects to spread out in a wide cone around its path. Just as a piece of paper falls as fast as a chunk of lead in vacuum, so such light sheets of aluminum or metalized balloons travel as fast as the ICBM in empty space. Not until they hit the top of the atmosphere do they slow down in relation to the ICBM, but by then the ICBM has only about a quarter of a minute to go to its target, after a flight of somewhat more than a quarter of an hour.

The counter-counter-countermeasure is to develop a short-range ABM that takes off so fast as to be able to intercept the ICBM in these last few seconds. In the United States this weapon is called Sprint. It carries a relatively small nuclear warhead, presumably about one kiloton and perhaps even less, only about one twentieth of the yield of the Hiroshima bomb. This is a small enough yield that if exploded perhaps ten miles above a city, it would do little direct harm other than blind people who happen to be looking at it. A possible counter-counter-counter-countermeasure would be for the attacker to arrange its ICBM so that it would detonate on interception. It is apt to be powerful enough to do a great deal of damage from an altitude of ten miles or so, though presumably less than it would do at an optimum altitude of two or three miles if it had not been intercepted.

The United States long-range ABM is called Spartan. It is said to have a warhead of about five megatons (in its original version) and a kill radius for ICBMs of presumably several miles (some say as much as tens of miles). The kill mechanism above the atmosphere is quite different from that in the atmosphere. In the vacuum of free space, x-rays propagate freely and evaporate material from the surface of the warhead so suddenly that the recoil creates a shock wave penetrating into it. Within the atmosphere, the kill mechanism used by Sprints presumably depends on a burst of neutrons causing enough fission to make excessive heat. X-rays are absorbed in air and are not of much use in the atmosphere.

The countermeasure to all this is MIRV, the Multiple Independently Targeted Reentry Vehicle. What this means is that one ICBM carries several warheads, each shielded to be able to reenter the atmosphere without burning up, and each of which may be aimed at a separate target. This is a way of making extremely many incoming warheads, so many that even a very sophisticated ABM system could not hope to handle them even if they could be distinguished from the chaff and other "penetration aids" in the last few seconds.

The transfer to dependence on MIRVs and thus greater numbers of warheads does not mean increased total yield or even increased area of destruction. On the contrary, the need to carry extra thrust and control machinery to guide the warheads independently means that less of the rocket payload may be devoted to nuclear explosives. Instead of one three-megaton warhead one might have ten warheads with perhaps only fifty kilotons each, for example. This would mean a reduction to one-sixth of the total yield of the more powerful warhead and a total area of destruction about two-thirds as great. However, this would be considered a military gain (even if there is no defense to be penetrated), because there are few targets worthy of the more powerful warhead, and the destroyed area can be distributed more effectively over a number of smaller targets. For example, in principle, it is possible for one missile with sufficient accuracy to destroy several missiles buried in the ground in hardened silos.

There are of course measures and countermeasures in other categories of weaponry. One goes under the name of antisubmarine warfare (ASW). The vastness and special characteristics of the sea, however, have thwarted ASW's effectiveness for many years. In particular, the boundaries between layers of different temperature and density of seawater reflect sound in such a way that it has not been found possible to detect submarines at great distances, particularly those powered by nuclear reactors so that they need not surface. It thus seems very unlikely that an appreciable fraction of submarines could be suddenly destroyed, and great confidence may be placed in the nuclear submarine as a secure base for launching the long-range missiles. If MIRVs should make ICBMs in their protective silos on land obsolete in the future, it is possible that submarine-launched missiles may tend to become the mainstay of deterrence, though doubtless the Army and Air Force would continue to press for alternatives to this dependence on the Navy.

With the mounting of countermeasure against countermeasure it appears as though the arms race could go on and on forever, with ever increasing technical sophistication, unless disaster intervenes. Unless somehow political restraint can be exercised, there seems to be no natural end short of the actual use of the weapons being deployed. This is particularly true as deployment spreads to many countries.

TESTING OF NUCLEAR EXPLOSIVES
Developing weapons of various kinds of course involves testing some of them. Armies and navies have long had their proving grounds or test-firing ranges not only to test the performance of new weapons but

to be sure on a sample basis that old ones are still working and to practice firing them and measure their destructive effects. Something as new and different as nuclear weapons of course had to be tested (although the one dropped on Hiroshima on August 7, 1945, was of the gun-assembly type that had never been tested).

The first nuclear explosive test was of the original 20-kT implosion bomb at the top of a 100-foot tower in the desert near Alamogordo, south of Albuquerque in New Mexico, not far from the remote Los Alamos laboratory where the first A-bombs were developed and designed in two and a half years of intensive wartime effort (Fig. 8–6). The fireball touched the ground and left a radioactive glaze on the surface. The debris was swept up in a mushroom cloud that was seen by an airborne observer when it reached its full height of 35,000 feet, to head eastward although the lower winds were northward. This was a first example of how everything in these tests does not always go just as might be expected. In the hours just after the predawn test, some small settlements to the east and northeast were evacuated to avoid radiation.

The Alamogordo test was the most significant of all nuclear tests. It was the culmination of six years of innovative effort since the discovery of fission in late 1938, effort to find out whether nature had so arranged nuclei that an explosive chain reaction is possible. This test gave the final answer, an answer that was felt by some of those participating to be ominous for mankind.

Since that time nuclear bombs have been tested both in the atmosphere and underground. The underground method was explored with a single explosion in Nevada in 1957, approximately 2 kT a thousand feet underground, and several more in late 1958 following a Geneva conference discussing a test ban. However, tests were generally carried out in the atmosphere until 1963 when the partial test ban treaty forbad this.

There have been many kinds of atmospheric tests, some in the vastness of the South Pacific, some in Nevada, some in remote parts of Siberia and the Arctic. One of the first after Alamogordo was of an A-bomb exploded underwater near a fleet of nearly expendable Navy vessels, to test their survivability. It raised a tremendous column of seawater and drenched ships with radioactive rain.

Most United States tests of A-bombs have been carried out in

Fig. 8-6 The present laboratory at Los Alamos, New Mexico, the mesa site where the first A-bombs were made.

Nevada, some on towers, some air drops. During the last half of the 1950s they were mostly quite "small" by nuclear standards as emphasis turned to developing tactical nuclear weapons. They ranged from 80 kilotons down to one ton (ton, not kiloton). Several were in the 100-ton range and many in the 1–20 kT range.

The era of thermonuclear explosions began with the "Mike" shot at Eniwetok in the South Pacific in November, 1952 (in the last days of the Truman administration). It was an awkward experimental device with refrigeration to hold the deuterium and tritium in place, a true hydrogen explosion but a tremendous one, the first to be measured in megatons. Its mushroom cloud was 25 miles high.

All this time the United States was keeping track of the progress of Soviet testing in Siberia by flying aircraft over the Pacific and western United States carrying air filters to collect radioactive debris. The next megaton shot was a Soviet one and was detected this way, the following

August (1953), and was observed to have involved lithium. So the Russians must have already exploited the lithium-deuteride idea that was still quite new to United States workers as the way to make a practical thermonuclear bomb, not just an experimental device.

The first United States test of such a bomb was the "Bravo" shot on the surface of an atoll at Bikini in the Marshall Islands the following March (1954). This shot was remarkable both for its success and for its unanticipated surprises. There is even some cause to wonder how responsible workers in good conscience could have set it off knowing so poorly what would happen. The central part of the atoll vanished, and in its place was left a deep crater filled with water. This was expected. But the yield was apparently two or three times greater than expected, 15 megatons rather than something like 6. (Its power so impressed the public that the atoll gave its name to an impressive garment from which the central part vanished.) The bomb was detonated quickly in hope of getting ahead of a shift of wind, but the shifting wind took the fallout toward the island of Rongelap. At sea nearby radiation levels were observed to rise, and the inhabitants were evacuated on an unplanned emergency basis from the inhabited south end of the island. Many of them developed severe skin burns and other injuries, and by now many of the adults and almost all of the children have contracted cancer. On the luckily uninhabited north end of the island, not far away, the radiation level was lethal.

At sea more than a thousand miles to the west, the Japanese fishing trawler *Lucky Dragon* was showered with a white ash that made the skin of the 17 crewmen itch and made some of them sick. When they put back to port in Japan, their fish catch was confiscated as radioactive, there was a great scare on the fish market in that fish-eating country already sensitive to radiation injury, and the crewmen were hospitalized. The radio operator died.[8]

In these above-ground tests the energy of the bomb is dissipated in the atmosphere in various ways: the shock wave, immediate radiation, and heat that carries the debris high in the air. Some of the radioactivity descends soon, in the next hours and days, and some not until years later.

In underground tests the mechanism of energy dissipation is very different. Normally the tests are so deep underground, typically a thousand feet or more, that the tremendous pressure developed is contained by the weight of the earth above. Much of the energy goes into vaporizing the surrounding rock as it compresses it to make an underground cavity. Some of the energy is carried away by a wave that is at first a shock wave and then subsides into a compression seismic wave that carries out to great distances. The part of the seismic wave that starts along the surface dies out, typically after about 500 miles, and the part that starts out obliquely downward bends up to reach the surface at considerably greater distances, typically 1000 miles. This leaves a "skip zone" in between where little or no signal is observed in an instrument called a seismograph that records a trace showing slight shaking of the earth.

Seismographs are normally set up by geologists to observe tremors from distant earthquakes. An earthquake involves faulting, or slipping of one section of rock past another, and sends out a wave that starts as a compression wave in two directions, for example north and south, and as a rarefaction east and west. By contrast with this, the seismic wave from a big explosion starts out as a compression in all directions. Thus it is in principle possible to distinguish between them by observation of different traces on the seismograph. In practice this requires several appropriately placed seismographs.

There are several hundred earthquakes a year making signals about as strong as a 5-kiloton underground explosion. With the state of the art in 1958, it was estimated (by an international conference of experts at Geneva) that 90 percent of these earthquakes could be distinguished from bomb tests by a system of seismographs placed 600 to 1000 miles apart over the land surface of the earth. Bigger earthquakes equivalent to a 20-kiloton test, for example, are much rarer and, with their stronger signals, easier to discriminate.

Since that time, with some interest in the possibility of a controlled ban on underground tests, great technical improvements have been made in the seismographic art. The limitation on a single seismograph is the local underground "noise," from passing trucks and the like. By electronic correlation of the signals from many seismographs in a widely dispersed array, it is possible to analyze the passing waves from various directions and to filter out the local noise. A separate improvement is gained by putting each seismograph into a hole drilled hundreds of feet into the ground, like an oil well. Seismographs have been specially designed for such installation (Fig. 8–7), and a large array has

Fig. 8–7 An especially designed seismograph being lowered into a deep hole.

been set up in Montana and tested with encouraging results. [10,11,12]

Most of the energy in the underground blast goes into vaporizing and deforming the rock nearby, leaving a cavity 100 feet or so in diameter (depending on the power) that is at first spherical and lined with a glassy substance formed from the condensing vapor and containing most of the fission products and other radioactivity. Within a few minutes after the shot, the roof of the cavity caves in, leaving a higher space until the cavity becomes cylindrical, with a height several times its diameter. It is mostly filled with rubble, with the highly radioactive glass at the bottom.

Altogether the United States has carried out underground nuclear tests ranging in energy release from a few tons all the way up to its two largest of 1 megaton and 5 megatons. Almost all have been at the Nevada proving grounds, but the two largest were more remote, on Amchitka in the Aleutian chain of islands extending out from the tongue of Alaska toward Siberia. The depth of such shots is scaled according to the power of the blast, calculated so the weight of earth should contain the pressure. The 5 megaton shot was over a mile deep, about 6000 feet. (Drilling the shaft cost about $50 million.) The largest of the Russian underground tests carried out in Siberia is said to have been about 6 megatons.

One might think that much of the desired information about a weapon might be hidden in a test underground, but apparently this is not the case. The original 1957 underground test was carried out in Nevada, as a safer way to test, by the Livermore Laboratory near San Francisco, a competitor of Los Alamos, much involved in testing. In spite of the great expense of preparing the deep hole, it was thought that it might in some cases also be cheaper for the United States, in comparison with the large test expeditions to the South Pacific, but not for the Russians. In comparison with surface tests on land it is expensive. Information about the explosive yield is of course easy to obtain underground from the intensity of the shock wave, but also other effects can be explored by special instrumentation.

REVIEW QUESTIONS AND PROBLEMS

1. Why does the assembly or compression of the fissile material in an A-bomb have to be fast?

2. What type of sudden assembly is usually used in an A-bomb?

3. What features of the A-bomb are useful in detonating an H-bomb?

4. How would the amount of fission products produced by a 200 kT A-bomb compare with that from a 20 kT A-bomb?

5. If a 10 kT A-bomb were used as the trigger of a 1 MT H-bomb,

how does the production of fission products per kiloton of explosive yield compare with this fission-products-per-kiloton ratio for a 20 kT A-bomb?

6. How does first-strike capability differ from second-strike capability? Which requires the greater disparity of forces between the two sides considered?

7. What feature of an H-bomb makes it possible to dress it up to make what is called a "dirty" bomb out of it?

8. In clear weather the ratio of firestorm damage to blast damage would be expected to be greatest for which of these bombs: a 20 kT A-bomb, a 500 kT H-bomb, or a 5 MT H-bomb?

9. What is considered to be the greatest danger of use of tactical nuclear weapons in presumably limited war?

10. Has the United States given tactical nuclear weapons into the possession of its NATO allies?

11. Have tactical nuclear weapons been used in warfare?

12. About how does the number of deliverable submarine-based missiles compare with the number of ICBMs in the United States strategic forces?

13. Describe at least one method for penetrating ABM defenses.

14. Does the area destroyed tend to increase or decrease as one goes to the multiple warheads of MIRV?

CHAPTER 9 Constraints on the Arms Race

DISARMAMENT ASPIRATIONS AND THE CLIMATE OF SUSPICION

Since the inception of the nuclear age it has been apparent to many thinkers that civilization would find itself in the terrible bind of a perpetual arms race with a high risk of nuclear war unless some systematic political restraint could be introduced. "One world or none" was a theme much discussed about 1945. The idea was not so much that there must be a world government with wide powers as that the world must be unified at least so far as the control of nuclear energy is concerned, in order to avoid a disastrous arms race.[1] Variations of the theme discussed the need for world control of all capabilities of making war or at least for a world military force more powerful than any national force, in order that no conventional war would break out that might through clandestine technical operations grow into a nuclear war.

The word "disarmament" had inherited a bad name from before World War II largely because the Naval Limitations Treaty of 1927, which limited the numbers of battleships retained by the major naval powers, had failed to prevent World War II. However, that treaty concerned only one element of national power that was losing its importance after the advent of air power. The feeling was that nuclear disarmament, involving the most decisive of weapons of unprecedented destructive power, would be much more influential in the avoidance of war, giving nations a much stronger incentive to enter into and abide by disarmament agreements.

Ever since the first discussions of nuclear arms limitation there has persisted in the United States a widely held opinion that it is futile or even dangerous to talk about arms limitation or disarmament because the Soviets have declared their intention to dominate the world and one cannot trust them in any agreements. This opinion has been so widely promoted through the years that the grounds for it need not be repeated here. Insofar as it is based on bellicose speeches, of which there have been many, it should be noted that there is a symmetry here; a selected collection of speeches by American military and civilian officials can be made to look just about as belligerent from the other side as the Soviet speeches do from ours. Among the purposes served by such speeches are fund-raising for armaments on our side and toleration of a lack of consumer goods on the Soviet side. It would be desirable if the average citizen, before deprecating arms control, could appreciate that practically all arms-limitation ideas propose to maintain dependable forces amply able to deter unprovoked attack and if he could study the history of disarmament initiatives enough to appreciate where the difficulties lie.

The first feeble move toward nuclear arms restraint was the Franck Report, submitted in the spring of 1945 by a group of scientists at the University of Chicago who were then among the few hundred people in the United States who knew about the prospective A-bomb. (See Appendix 16.) They were members of the effort that went under the purposely misleading code name "Manhattan Project," contributing to the making of the bomb. The Franck Report, named for one of the Chicago scientists, James Franck, recommended that the bomb should not be used against Japanese cities but rather in a demonstration for the Japanese military and representatives of many nations to see, destroying at one blow some uninhabited region (perhaps a Japanese island). It was considered in the report that such a demonstration should be sufficient to convince the Japanese to surrender. The report emphasized that the first use of the bomb was not only an opportunity to end a bloody war then in progress but was the beginning of a new and dangerous atomic age, and that restraint in our manner of entering into it would influence the prospect of organized restraint among nations to avoid the worst dangers of an atomic arms race.

This idea of demonstrating the power of our new weapon to the Japanese enemy without widespread carnage was considered seriously by committees influencing Secretary of War Henry Stimson, and the consensus in the committees and among many of the scientists consulted—who were uninformed about the secret Japanese peace overtures then known only to top officials—was that the remarkable tenacity of the Japanese military machine would not be broken by a remote demonstration. If this was true, despite the peace overtures, then the further influential argument is valid that many more Japanese as well as American lives were saved by avoiding the massive land attack on Japan to end the war than were lost by the atomic attacks on Japanese cities. They did bring the war to a prompt end. The Franck report itself, pleading for the demonstration approach, never got far enough through channels to reach President Truman before he, soon after taking office and learning about the bomb, made the fateful decision that resulted in the destruction of Hiroshima and Nagasaki and demonstrated to the world the problem it faces in the nuclear age. Even many of those who favored the demonstration on a city questioned whether a second tragic demonstration only three days after the first, a gargantuan one-two punch, was justifiable.

ARMS-CONTROL NEGOTIATIONS

Once World War II was quickly brought to an end with the use of two atomic bombs, the world of diplomacy was faced with the challenging question whether the technical innovation that had led to this unprecedented and dangerous new order of magnitude of destructive power could be matched by political innovation adequate to prevent its further use for human carnage. Politics had for centuries pitted nation against nation in seeking special advantage, and it was doubtful whether even the demonstrated enormity of the nuclear threat could lift peoples' thoughts away from old habits considered practical to imaginative and seemingly necessary new ways considered idealistic.

The Acheson-Lilienthal Proposal

The next initiative towards restraint was the Acheson-Lilienthal plan, which later took on the name of Bernard Baruch. A committee set up by Secretary of State Dean Acheson was chaired by David Lilienthal and included Robert Oppenheimer, an important theoretical physicist turned wartime chief of the Los Alamos Laboratory that produced the bomb. It is said that most of the ideas of the report were his, nurtured by the ferment of discussions among scientists of the project in the last months before and just after the bomb was used. The report proposed that there be set up a world authority with complete control over the world's supply of uranium and with sole possession of the capability of making atomic bombs (whether the capability of making them was used or not) and that it be established as the world leader in research in all nuclear matters. Being on the forefront of knowledge, this authority would thus know what to look for in carrying out its charge to prevent any clandestine dangerous nuclear activity.

In retrospect, the plan should be considered generous on the part of the United States in that it involved giving up an American (plus British) monopoly on A-bomb technology in about the only way that could be devised to satisfy reasonable demands of military safety. From the Russian point of view, however, it might be seen as not so favorable because it required inspection of Russian uranium mines before surrender of American know-how. Under Stalin, the U.S.S.R. was not yet ready for any such advanced ideas. Neither was the United States, as evidenced by the fact that the plan was submitted before the United Nations in modified form by a conservative elder statesman, Bernard Baruch, who placed on it restrictions and a tone of presentation calculated to guarantee Russian rejection. (Reference 2, page 191.) As presented, it became known as the Baruch Plan.

For several years American diplomacy stood on its laurels, having made this supposedly generous proposal. Meanwhile the Russians went ahead and developed the A-bomb, which they first tested four years later, in 1949.

Secrecy, Inspection, and Asymmetry

After this there was a period when each side vied with the other in making disarmament proposals which sounded good superficially but were unacceptable to the other side. In the gulf between them lay the issue of inspection. The Russian Communists, like the czars before them, were and to a lesser extent still are very leery of intrusion by foreigners. For years the Americans accused the Russians of wanting disarmament without inspection (as verification of compliance), while the Russians accused the Americans of wanting inspection without disarmament. The American proposals, with understandable but perhaps excessive caution, demanded setting up complete systems of inspection before starting to reduce the number of arms.

In those early days of the nuclear age there was an asymmetry in the military balance between East and West. World War II ended with the American military machine the strongest the world had ever known, but on the basis of United States monopoly of the A-bomb it could be withdrawn from Europe and mostly demobilized in spite of Western distrust of Soviet intentions. This distrust arose largely from the fact that the U.S.S.R. took over as a permanent sphere of influence the countries of eastern Europe that it was occupying militarily at the time of German collapse, though this had not been the western intention in the Yalta agreements arranging for this temporary occupation. It seemed to the West an indication of Communist expansionism that might encompass all Europe if not militarily opposed, though historically it was quite consistent with the old practice of establishing a *cordon sanitaire* or buffer zone of friendly states around any great power.

The asymmetry consisted, then, of United States superiority (or initially monopoly) in nuclear weapons and a rather pronounced Soviet superiority in land armies, including thousands of heavy tanks suitable for invasion. This complicated discussions of possible disarmament. The Soviets at first wanted to start with complete elimination of nuclear weapons—"ban the bomb"—which was obviously seen from the West as leaving western Europe almost defenseless. The West wanted to establish specific quotas for the size of the various national military establishments (about $1\frac{1}{2}$ million men each for the U.S. and the U.S.S.R., $\frac{3}{4}$ million each for France and England) as a part of the price of balanced reductions in atomic stockpiles. It also wanted complete and essentially continuous inspection of all important factories. As the tempo of negotiations in the U.N. disarmament committee stepped up with the advent of the H-bomb in 1952 and the end of the Korean war in 1953, these proposals were spelled out by Britain and France, with the United States concurring. The Russians held out for percentage cuts in armies (that would have left theirs biggest) and rather limited inspection of atomic installations. These understandable differences illustrate how hard it is to make a disarmament agreement that will satisfy the military demands for security by suspicious adversaries. All along there have been some highly placed men aware of the danger of a long nuclear arms race and sincerely trying to find ways to end it, but they did not succeed in resolving such differences![2]

An Overture and a Retraction

In 1952 Stalin died and after an interim Khrushchev took his place as Soviet leader. That there was a sudden change of Soviet policy for the better was evidenced by the sudden resolution of the Austrian peace treaty, freeing that country from the shackles of four-power occupation. In May, 1955, Soviet disarmament policy did a sudden about-face, coming really very close to what the West had been proposing. It accepted the numerical ceilings on armies and proposed verification on a "permanent basis" by inspectors who "within the limits of the control functions they exercise, would have unhindered access at any time to all objects of control."

There were still problems about searching for other objects of control and about abandonment of United States military bases in countries like Turkey near the U.S.S.R., but the change was so drastic as to seem a real opportunity to negotiate a meaningful and mutually advantageous disarmament treaty if the West had really wanted to.[2,5] After two months of consideration, it became apparent that the West did not want to, even on the terms it had been proposing. President Eisenhower expressed doubts that the American people were ready to accept inspection on the ground but proposed instead "open skies," unrestricted photo reconnaissance of each country by the other.

All previous United States proposals were then explicitly withdrawn, with the explanation that these would require more men to man United States bases and that so much nuclear material had already been produced as to make it technically impossible to be sure that stockpiles were cut back all the way to zero. If we had wanted substantial nuclear disarmament, the goal of cutting back to zero could probably have been negotiated away. Basically, the role of conventional military thinking in the formation of United States foreign policy was too strong to permit concessions in the matter of foreign bases for the sake of the far more important objective of nuclear disarmament. Most of the bases were considered essentially obsolete about a decade later. It might have been necessary to negotiate a united and neutral Germany on the Austrian pattern, which would have put a crimp in NATO but would also have avoided some of the need for it.

Of the Russian about-face, the intrepid French negotiator Jules Moch said, "The whole thing looks too good to be true." Of the American retreat from its position, Philip Noel-Baker in *The Arms*

Race,[2] said, "The detailed plan for inspection and control, all the other proposals urged with such vigor and persistence only three months before—all were withdrawn." The fact that the proposals were withdrawn is perhaps only a sign that the United States government had not bothered to do its homework on disarmament as long as it thought the U.S.S.R. was not interested in any realistic propositions. When faced with the real possibility of agreement, it wanted to be more realistic about working out the details. The real indictment of United States policy, the real indication that military advice was always ready to pull the rug out from under the negotiators if there should be "danger" of agreement, is found in President Eisenhower's "open skies" proposal. In effect, this said to the Russian leaders, "Now that you seem ready to agree to the kinds of inspection we were demanding, but to which we are no longer sure we want to submit ourselves, we hereby up the ante and demand a more intrusive type of inspection."

Arms Control and Disarmament Agency

One trouble throughout the fifties was that governments—or at least the United States government—seemed to have no agencies with enough capability to explore and assess the whole realm of ideas about disarmament in advance of negotiation. There was a small disarmament staff in the Department of State, but for the most part disarmament ideas were evolved in the forefront of negotiation and usually by a negotiating delegation of prominent people from outside government and with little previous concern for disarmament, assembled hastily a few weeks before each disarmament conference.

Some scientists and others as early as 1951 pointed out the need for a sort of national laboratory for armament-restraint ideas and assessments, but it was not until 1961 that the United States Arms Control and Disarmament Agency (ACDA) was established to perform at least part of this function and in addition to represent the government at the negotiating table. Though housed with the Department of State, the ACDA is essentially an independent agency, its director reporting directly to the President. It is a valuable source of initiative for arms-limiting proposals within our government, but its negotiating positions must be approved by several other authorities, and it is such a minuscule agency compared with the Department of Defense that it usually seems to have insufficient influence on final decisions.[8,23]

Nuclear Test-Ban Talk and Arms Build-up

But to return to the fifties. It turned out, then, that the British-French-United States proposals that evolved about 1954 seemed to have been largely window dressing so far as the United States was concerned. Nevertheless, the negotiators did continue to try to develop ideas and hammer out differences. The Russian conciliatory proposal of May, 1955, had emphasized control of conventional arms (as had the discussions preceding it), and later negotiations worked out improvements on the nuclear side.

The idea of an agreed ban on all testing of nuclear weapons came to take a more and more important part in this. At about the same time the idea of a test ban sprang into political prominence in the United States through the initiative of Adlai Stevenson. It had been promoted some time before by scientists as a way of slowing down the development of ever more terrible nuclear weapons,[9] but in his campaign for the presidency in 1956 Stevenson publicized the health hazards from radioactive fallout as the main reason for a test ban. This was about a year after the significance of strontium-90 in milk had been discovered.

Unable to attain agreement on more ambitious disarmament measures, in 1958 and 1959 the negotiations centered on the possibility of a nuclear test-ban treaty. As the negotiations, after further setbacks in 1959, again began to look hopeful, a summit conference between heads of state was arranged where some progress might have been made. However, this was cancelled after the capture by the U.S.S.R. of the U-2 high-altitude photo reconnaissance spy plane, which disclosed that the Central Intelligence Agency of the United States was actually practicing "open skies" on the sly.

The Kremlin now apparently despaired of doing any disarmament business with President Eisenhower at the end of his administration but gave indications that it might with President Kennedy in 1961. Eisenhower left office having built up and authorized the completion of an enormous nuclear striking force, over 600 ICBMs, about 100 missiles in 6 nuclear submarines, and over 3000 bombers capable of attacking the U.S.S.R. Altogether this was roughly 10 times the Soviet's nuclear striking power, and it has been estimated that it was about 10 times as much as needed to make uninhabitable all the inhabited part of the U.S.S.R.

Yet John Kennedy came into the presidency eloquent in the cause of nuclear disarmament and at the same time accusing the Eisenhower administration of having endangered the country by letting the Russians get far ahead of us, creating a mythical "missile gap" in their favor. He sponsored the creation of the ACDA, which had real value, and gave other indications of favoring arms limitations while presiding over a further enormous missile build-up, to the level of 1000 Minutemen on land and 656 Polaris missiles in 41 nuclear submarines. Much later, even as late as 1968 after the Russians had long been trying to catch up part way, his Secretary of Defense, Robert McNamara, then serving under President Johnson, was able to assure the nation we had 4 times as many deliverable nuclear warheads as the Russians had.

As an eloquent gesture toward disarmament hopes, Kennedy designated John J. McCloy as his personal representative to meet with Khrushchev's representative, Ambassador Valerian A. Zorin, and work out the outline for an agreement on disarmament. In a signed document submitted to the United Nations in September, 1961, as a basis for more detailed negotiations, the two nations agreed to the principle of abolishing weapons of mass destruction, reducing conventional forces to a minimum, and establishing a veto-free international central agency giving inspectors unrestricted access for effective verification. On this ray of hope the patient work of a decade of disarmament hassles seemed to have paid off in bringing the stated viewpoints of the two sides together. Yet the pressures for continued arming seem to have been stronger than these pressures for restraint, and the McCloy-Zorin agreement was soon forgotten.

Momentum of the Arms Race

It has been standard practice for our national leaders, and the leaders of other nations as well, to try to please everybody by saying one thing and doing another. The voluminous "Pentagon Papers" on the origin of the South-east Asia involvement, classified "secret" and released in June, 1971, through the initiative of *The New York Times*, reveal the extent to which successive presidents become enmeshed in deceit about foreign policy formed primarily by the military and "intelligence" branches of government. But it is hard to understand why a man of Kennedy's good will and broad sources of advice could not have given stronger support to the efforts to tame the nuclear arms race, even while presiding over increasing nonnuclear military involvement.

President Eisenhower, apparently appalled by the missile build-up that had gone on in his name and disappointed that he could do so little to establish a basis for permanent peace, left office in 1961 with that famous valedictory admonition, "In the councils of government we must guard against the acquisition of unwarranted influence, whether sought or unsought, by the military-industrial complex. The potential for the disastrous rise of misplaced power exists and will persist." It seems that men in high places can thus unburden their souls only while stepping down.   About eight years later, Secretary of Defense McNamara, who had played an important role both in promoting military procurement and in introducing some restraint into it in the Kennedy-Johnson era, made a similar valedictory statement. In a speech on September 18, 1967, he warned against the futility and danger of deploying a large ABM system and at the same time announced the deployment of a relatively small system that would almost inevitably grow into a large system. He warned, "There is a kind of mad momentum intrinsic to the development of all new nuclear weaponry. If a weapon system works—and works well—there is a strong pressure from many directions to procure and deploy the weapon out of all proportion to the prudent level required." (See Appendix 18.)

This "unwarranted influence" or "mad momentum" is augmented by the vicious interservice rivalry within the Department of Defense, the separate influence of the Army, the Navy, and the Air Force on Congress and the executive branch to give it at least as much as it gives the others. All this helps explain why Kennedy, hoping to be a man of peace, built up such a large degree of "overkill." Arthur Schlesinger, Jr., in his book about the inner councils of the Kennedy administration, says:

[McNamara] was already engaged in a bitter fight with the Air Force over his effort to disengage from the B-70, a costly high-altitude manned bomber rendered obsolescent by the improvement in Soviet ground-to-air missiles. After cutting down the original Air Force missile demands considerably, he perhaps felt that he could not do more without risking public conflict with the Joint Chiefs and the vociferous B-70 lobby in Congress. As a result, the President went along with the policy of multiplying Polaris and Minuteman missiles.*

What chance sanity toward disarmament in such a maelstrom?

---

* Arthur Schlesinger, Jr., *A Thousand Days*, New York, Simon & Schuster, 1965, p. 500.

TRANSITION TO DISARMAMENT

To turn from the history to the substance of nuclear disarmament problems, we must realize first that the Soviets even more than the West looked on military secrecy as an important element of military strength. Before the days of reconnaissance satellites, secrecy of the locations of missile launching sites could prevent the West, with much larger numbers of missiles, from wiping out Soviet missiles, for example. With the less open society of the U.S.S.R. such secrets could be kept. Satellites don't tell everything, and there is still military value in secrecy.

In principle, the technical problem of disarmament is to devise a plan whereby suspicious adversaries may agree to reduce the level of armaments while providing each other with assurance of compliance so that neither is taking a chance that the other, by merely pretending to comply, will achieve a stronger military position. The dilemma throughout the fifties was that inspection and military secrecy seem to be incompatible: inspection penetrates secrecy. The Soviets quite logically proposed that if the West would agree to general and complete disarmament, they would agree to admit inspectors with free access everywhere, for there would then be no military secrets to hide. This unresolved dilemma for partial disarmament could be used as an easy excuse for the lack of political progress.

The behind-the-scenes concern about the problem did evolve a remarkably simple solution that could have been the basis for cautious and effective progress in disarmament if the factions desiring this on the two sides had been more influential. At a meeting in Moscow in 1960, one of a series of conferences of scientists on world affairs that go under the name of the "Pugwash" movement, it turned out that this solution had been invented independently in the U.S. and the U.S.S.R., here by Louis Sohn and in the U.S.S.R. by A. P. Alexandrov. Sohn calls it "territorial disarmament" but a more descriptive name is "region-by-region disarmament."[7] While the problems would be quite different today if there were diplomatic interest in disarmament, elements of this solution would still be applicable and it is instructive to see how detailed ideas can in principle help solve a general political problem.

The proposal contains three special features. First, instead of disarming gradually by categories of weapons, it is proposed to disarm gradually by regions within each country. The proposal seeks to disclose secrets such as missile sites and introduce inspectors simultaneously in one region as it is disarmed. This eliminates only a fraction of the national military power at one stage and is thus an acceptably gradual approach to disarmament. The regions are chosen to be of equal military value, and it is not known in advance which is to be disarmed until each stage commences. Second, there is a preliminary inventory of all military installations in all the various regions, without disclosure of their exact locations. Third, at the time a region is selected to be disarmed, a detailed inventory must be submitted agreeing with the preliminary one and disclosing locations of equipment within that region. This provides an opportunity, by thorough inspection of the region, to verify the original inventory on a sampling basis and to establish confidence.

To explain a bit more fully, the region-by-region plan starts out by having each nation divide itself into, let us say, six regions, which it considers to be of equal military value as suggested in Fig. 9–1. (If it divides them unequally, it may be left short in the late stages.) A complete inventory of arms and military installations within each region, but without their exact location, is submitted to an international control authority. Inspectors may be spread thinly on the borders between the various regions in advance. Then, on a given day, the first-stage regions to be disarmed are chosen in an unpredictable way, either by lot or by choice of the other side, and the international inspectors are quickly concentrated on the borders and airports of that region to seal off arms transport. Detailed inventories giving locations are then submitted for these chosen regions and the inspectors disarm them.

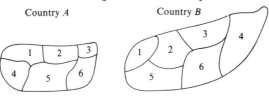

Fig. 9–1 Typical division into regions in the region-by-region disarmament plan.

The important point is that the inspectors then have a year or so to search the region thoroughly with complete access, to verify that there are no hidden arms. This constitutes a random sampling of the previously submitted inventory, and on a sampling basis verifies that it is fairly dependable. Thus some confidence is developed, each side has less distrust of the other, and as the successive steps proceed, region by region, the sampling becomes more adequate and confidence is further increased. Thus disarmament and inspection spread together gradually over each country, region by region. (See Fig. 9–2.) When the time comes to disarm the last region, each side has strong reason to believe that the inventory submitted by the other is accurate, and ideally each should be able to trust the completeness of the sudden disarmament of the last region in its turn. Two important features are, first, that a nation would not know which region was to be the last, and so could not prepare it in advance, and second, that there must be some sort of inspection of production facilities (which will not disclose geographic military secrets like missile sites) to prevent production of important weapons in the last region while the next-to-last is being disarmed.

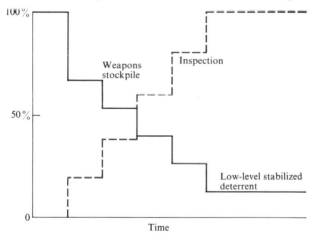

Fig. 9–2 Diagram showing how the amount of inspection increases gradually as the number of nuclear weapons decreases in the region-by-region disarmament plan.

This type of procedure is adaptable to various disarmament goals. With present world tensions a low-level stabilized deterrent would be a more realistic goal than complete disarmament at the end of the last stage. The arms race has produced levels of nuclear striking power far greater than needed for effective deterrence. If each side had a hundred

ICBMs and knew that the other side had no more, this should be ample. Even General Taylor, Secretary of the Army early in the Kennedy days, considered such a level enough without any agreement. (One might better say fifty or twenty missiles on each side.) The purpose of retaining a low-level deterrent (at least until something in the nature of world government evolves) is that it covers uncertainties. If one side by cheating could produce ten H-bombs on the sly, this could be a crucial threat if both sides were otherwise disarmed but would make little difference if each had fifty or a hundred missiles.

The point is not that a prospective nuclear war would be appreciably less terrible but that the missile system would be less accident-prone and the political situation would be less jitters-prone, so nuclear war would be less likely. In a calmer atmosphere without an arms race, the intelligence of mankind might devise a more satisfactory long-term riddance of the threat of nuclear terror. Whether or not that might happen, the low-level stabilized deterrent would be much safer than the arms race without terminal facilities.

ARMS CONTROL: THE NUCLEAR TEST BAN

As it became increasingly clear that the arms race had progressed so far that nuclear stockpiles were enormous and disarmament had proved elusive, emphasis turned to seeking methods of controlling the development and deployment of arms so as to slow down the arms race or to stop it. This is consistent with keeping some degree of disarmament as a future goal. The thought is that the arms race must at least be stopped before the trend of arms stockpiles can be turned back toward lower levels. A ban on nuclear testing is one method of arms control through the control of the development of more sophisticated nuclear weapons. Though Eisenhower had to oppose Stevenson's promotion of a test ban in the 1956 political campaign, he soon after supported the quest for a test ban for arms-control purposes.

In 1958 there was in Geneva an important conference of experts (mostly scientists) from East and West to explore the possibilities and outline a proposed treaty. The major technical problem arose from the fact that nuclear arms development, with some inconvenience and considerable extra expense, could be carried out deep underground. Then tests would not be easily detected far away by sampling radioactivity in the atmosphere, as can be done with air bursts.

However, underground bursts can be detected instead by observing seismic waves propagated within the earth itself. The difficulty is that detection requires rather detailed observations, preferably from not many thousands of miles away, to discriminate between an explosive burst and an earthquake. A system was worked out and proposed whereby there would be international access to a network of seismometer stations in each of the countries concerned, to give as reliable signals as possible. Also on-the-spot inspections would be permitted at least at some of the rather few places where the signals would indicate that there had been either a quake or a burst, but one could not be sure which. Since there are many very weak earthquakes, it is not practical to push the discrimination down to small nuclear bursts. The experts from both sides at Geneva agreed that the very small bursts would be relatively unimportant for weapons development, not worth breaking a treaty, since so much was known already about making weapons. The system proposed was elaborate enough to detect and discriminate bombs of about 20 kT or more detonated in small cavities deep underground.

Calm people considered it reasonable to ban all nuclear explosions by treaty, trusting that neither side would cheat but knowing that the strategic balance could not be upset in any case by a little undetected cheating. They considered that the advantages of slowing down the arms race by eliminating the enormously powerful tests was worth any small risk that one side might possibly gain a little by clandestine activity. But there were those in the United States, including scientists, who felt that this would be a betrayal of our national interests. Representatives of the Livermore nuclear weapons laboratory thought we had "lost our shirts" at Geneva and proceeded to invent the "big hole" to torpedo the proposal. They pointed out quite correctly that tests larger than 20 kT could evade discrimination if carried out in large underground cavities, but they went to the ridiculous extreme of crying out the danger that a cheater might test a MT-range weapon in an impractically enormous spherical underground cavity 800 feet in diameter. Such a cavity is many times larger than any ever excavated and over a hundred times the volume of the largest one the United States later found it practical to build (without any limitations of secrecy) to test the "muffling" of the seismic signal.[10] Through exaggeration of such fears, United States paranoia prevailed at a time when the United States had something like 10 times the nuclear striking power of the U.S.S.R., and the high hopes of 1958 for a complete nuclear test ban were watered down to a United States proposal for a ban of tests in the atmosphere but not underground.

In October, 1958, just after the conference of experts, the U.S. and the U.S.S.R. jointly declared a one-year moratorium on nuclear testing to clear the political atmosphere for negotiations. A year later it was extended for a few months, but on December 27 President Eisenhower formally announced that this country would consider itself no longer bound by the terms of the moratorium after January 1, 1960. Thus by our initiative the moratorium was declared dead. Yet when the U.S.S.R. suddenly resumed testing in the atmosphere in a big way in mid-1961, there was a cry in the United States that the U.S.S.R. had perfidiously broken the moratorium, and this was advanced by opponents of agreement as showing that an agreement could not be made with the U.S.S.R. Our officials, having been so naive as to expect that the test cessation would endure even without any agreement, seemed to be hiding their embarrassment behind a big misinformation hoax about Russia's "breaking" a moratorium agreement.

The ensuing two years were the period of the most active atmospheric testing of all and the U.S.S.R., having prepared for it in advance, carried out more tests than we did during the first few months of that period. The United States was interested in developing the technique of underground tests and started off with some of these. The most powerful test made was by the U.S.S.R., a 57 MT shot. This is more powerful than makes sense for a counter-city weapon (15 MT is usually considered a practical limit for such purposes) and was probably tested mainly for propaganda purposes, though in an all-out war bombs in the 100 MT range could be used to ignite thousands of square miles of forest fire. Not only the forests would thus be destroyed, but also agricultural land would be washed away by subsequent flash floods.

As the tests continued and the fallout mounted, President Kennedy kept pressing for a ban on tests in the air, underwater, and in outer space. Though this would be a good example of mutual restraint for mutual benefit in getting rid of fallout, it could be seen as militarily favoring us because we, more than the U.S.S.R., can afford the extra expense of testing underground. The U.S.S.R. insisted on a ban on all nuclear tests including those underground.

On May 10, 1963, Khrushchev suddenly performed an about-face. He was apparently really worried both about the danger of nuclear war and the drain of the arms race on the strained Soviet economy and seemed genuinely interested in obtaining agreements to stop the race. The momentous ideological split with China was over this issue. Khrushchev on that date accepted a bargaining defeat and agreed to the partial test ban proposed by Kennedy. He did so apparently in the hope that it would soon be followed by other arms-control agreements. In return he even asked for a nonaggression pact with NATO and the Soviet satellites, but Kennedy denied him this.

Kennedy hailed the "half a test ban" treaty of 1963 as a great stepping stone toward eliminating the scourge of nuclear war, but nothing further came of it as we proceeded with our rapid missile build-up. Khrushchev was apparently discredited at home for having given and not received. He fell from power within a year, the West having again lost an opportunity for initiative toward arms control.[20]

In the years since 1963, the United States has tested more megatonnage underground than it had above ground before the ban, about 350 tests. The treaty permits only those tests so completely contained underground that no radioactivity can be observed beyond national borders. About 20 United States tests in Nevada have "vented," and at least two have been officially reported as observed in Canada. Thus, after all our worry that the U.S.S.R. would cheat, we have violated the treaty and still go on carrying out tests that might vent, including bigger tests up into the multimegaton range. Such are the pressures from the promoters. It may be that both sides, perhaps by tacit agreement, are violating the treaty provisions with weak radioactivity across national borders, trusting that this will not get out of hand.

It is somewhat alarming to think how much radioactive pollution of the atmosphere might have been caused by continued atmospheric testing by now if there had been no partial test ban treaty. The treaty has served as an example of mutually beneficial restraint carried out by agreement between the two major powers, though the benefits have been slightly one-sided because of the underground test situation. In retrospect the treaty appears to have been an agreement that the ACDA under President Kennedy was permitted to make because the influential Department of Defense and AEC leaders had developed enough competence and confidence in underground testing to feel that the atmospheric test ban would not seriously hamper their operations (and, as we have said, would hamper thc Soviets more). Now that underground operations are becoming so powerful as to make it seem advantageous to break out of the restrictions, there is already pressure for modifying or abrogating the test ban formally as we continue mildly to break it, or at least bend it, informally. There is some danger that even this rare success of atomic diplomacy will be sacrificed to commercial, if not military, pressures and that the world will revert to an era of unrestricted atmospheric testing with the accompanying fallout and increased international tensions. Having been caught once unprepared, we now have a "readiness" program and could start atmospheric testing at a moment's notice.

On the other hand, it would seem to be in the interest of both nuclear giants to agree on a complete nuclear test ban. Much has changed since 1958 when an attempt to achieve it failed. The emphasis of development is no longer on tactical nuclear weapons, for which clandestine tests of very low power might have some significance, but rather on ABMS and MIRVs involving larger warheads and more sophisticated tests. The impressive technical development of means of detection and discrimination of seismic waves, involving the large arrays of instruments mentioned in Chapter 8, make it likely that satisfactory verification could be provided without the politically sticky need for on-site inspections. A test ban would strengthen a further agreement to ban ABMS and MIRVs. Even without further agreement, a complete test ban alone would slow down the development of ABMS and MIRVs and decrease the tendency towards reliance on these systems that destabilize deterrence. Finally, both as a tangible example of restraint by the "giants" and as an obstacle to independent weapons development, it would help to convince nonnuclear countries not to "go nuclear."

ABMs AND MIRVs AS OBSTACLES TO ARMS CONTROL

The missiles in most pressing need of arms control or simple restraint are the qualitatively new ABMs and MIRVs. Because they destabilize the deterrent by introducing great uncertainties in the estimate of its effectiveness, leaders on both sides should appreciate that they would be better off if neither side had them. The limitation on numbers of these weapons achieved in the 1972 SALT agreement (see page 242) indicates that this is to some extent appreciated, but the dynamics of the arms race seems still to push development of more advanced weapons within the permitted numbers.

Naively and at first sight, it seems as though antiballistic missiles should be a good thing, for they are purely defensive and, in the event of an attack of a given size and pattern, would save some lives to the extent that they function at all. The fallacy of this reasoning is that the ABMs influence the likelihood that war will break out. They also influence the size and pattern of the anticipated attack, all in such a way as to make the situation worse if we go ahead and install them than if we do not, in the opinion of most analysts.[13,15]

The introduction of great numbers of nuclear weapons into national arsenals leaves much of mankind living under a very real threat of terrific destruction in nuclear war. The threat does not always seem real only because it is too terrible to think about continually and because we have been prudent enough or lucky enough to have avoided it for over two decades. It would be good to get rid of that threat. Though the immediate aim is less ambitious, the ultimate aim of disarmament negotiations is to do just that and at the same time remove the causes of conventional wars. In view of the difficulties along the way, as a next-best goal it would be good to have a completely effective defense against all means of delivery of nuclear weapons. That would eliminate the threat. Thus one might wish for a completely effective ABM, though this would interdict only one means of delivery.

The trouble is that any foreseeable ABM system is very, very far from having any such capability.[17,25] Successful simulated interceptions have been carried out with one ABM against one ICBM, neither carrying nuclear warheads, under ideal test conditions when a special ABM crew was alerted for the event. Success in this case means that the ABM received a signal to detonate when it was close enough to the incoming ICBM that if it had carried a nuclear warhead it would theoretically have destroyed the ICBM. This meant that one radar had to handle one incoming missile at a preset time. Compared with this, the task of handling dozens or even hundreds of incoming missiles accompanied by all their penetration aids in a short time in an actual nuclear surprise attack after years of alert is so fantastically more complicated that most unbiased experts believe that no more than a small percentage of the attack could be intercepted. The computer coding problem alone is said to be far beyond anything that has yet been accomplished in this already sophisticated field. Complicated systems usually need testing and tuning before they work well, but for this system there is no realistic test short of the actual attack, so it must work perfectly the first time with a few minutes warning, after sitting idle for years and in spite of enough safety catches to avoid misfirings in routine maintenance.

Nevertheless in June, 1969, Congress under pressure from the Nixon administration authorized the actual deployment of an ABM system after years of development. At that time, on the basis of the technical misgivings discussed above as well as the political disadvantages of a perpetual arms race, six men were in the best position to have had access to all the technical details without having acquired departmental bias (past presidential scientific advisors and the current one, joined by a former Pentagon research chief) all opposed the deployment. Direct and insistent political pressure was brought to bear on senators by the Department of Defense, using as a lever the Department's expenditures in their respective constituencies. On the other side, the public hue and cry against the ABM was enormous and unprecedented, and had a nearly telling effect.

It is an interesting commentary on political pathways to note that the public concern arose not primarily from thoughts about international repercussions or financial waste or technical shortcomings but from worry about "H-bombs in the backyard" of the big cities. Late in the Johnson administration an ABM system known as Sentinel was proposed, using the big long-range Spartan ICBMs, intended for area defense over a region several hundred miles wide. These were located for convenience at the same site as the short-range Sprints intended to provide some local protection to the big cities. Local opposition aroused enough congressional concern that this system was not deployed. The Nixon administration the next year eliminated the political obstacle by proposing a smaller system (called Safeguard) with long-range missiles at only two centers near ICBM sites far from big cities. There was still strong public pressure to stop ABM. The famous 50–50 vote in the Senate in 1969 on an amendment deleting ABM funds was a single vote short of stopping the deployment, though such positions against military funding had seldom attracted more than two or three votes in the past.

At that time the Pentagon claim was that although an ABM system could not be effective against a massive Russian attack, it would protect

Fig. 9-3 The Conference of the Committee on Disarmament (CCD) in session at Geneva.

us against an attack by the Chinese when they might have a few missiles in the mid-1970s—if they should be so brash as to make a suicide attack against us, undeterred by our overwhelmingly massive force which we expect to deter the much better equipped Russians. The system was characterized as a small system to provide operational experience in deployment, not a first step to a big system. Yet within less than a year the administration was pointing to the need for defense against the Russians and pressing to enlarge the system.

Throughout most of its existence since June, 1945 (the month before the first atom bomb), the United Nations has had a disarmament commission or committee in which the problems of disarmament, mainly nuclear arms limitation, could be discussed and negotiated. From 1962 to 1969 it was called the Eighteen-Nation Disarmament Committee, ENDC, supposedly consisting of representatives of five NATO powers, five Warsaw-Pact powers, and eight nonaligned nations. France refused to participate (presumably because she could not be treated as equal to the nuclear giants in nuclear matters) so the ENDC had seventeen members. Since 1969 the same meetings have been known as the Conference of the Committee on Disarmament, CCD, (Fig. 9–3). Its main successes have been the negotiation of the partial test ban and the nonproliferation treaty. Other peripheral arms-limitation matters negotiated there have been the prohibition of nuclear weapons from the Antarctic and the sea bottom.

However, the prospect of an ABM-MIRV confrontation concerned the two nuclear giants too directly to be negotiated in so large a group. While a desire for ABMs and MIRVs seemed to dominate the Department of Defense's influence on United States policy, the Soviets (who had started the trend towards ABMs with their light deployment around Moscow) apparently recognized that the ABM in the long run would do more harm than good, perhaps introducing uncertainty into the adequacy of their deterrent against us. In the last months of the Johnson administration in 1968, they proposed bilateral negotiations between the United States and the U.S.S.R. to limit ABM deployment and perhaps also to put some limitations on offensive missiles. President-elect Nixon requested that our agreement to enter into talks be postponed for his administration to handle, since he would be responsible for the negotiations. After he took office there were months of delay before the Strategic Arms Limitation Talks (SALT) were allowed to begin. They have continued intermittently, alternating between Helsinki, Finland, and Vienna. For the first two years they made little progress, while testing and perfecting of MIRVs and initial deployment of both MIRVs and ABMs continued to make it more difficult to agree on means for stopping them.[21,22] (See Appendix 17.)

In the first year and a half of the talks there was not even agreement on what to negotiate about. It was expected that there should be some trade-offs between limitations on ABMs and on the offensive or strategic missiles with which the two sides threaten each other's homeland. While some of the details of the talks have been secret, the main bone of contention appears to have been over the determination of which missiles are strategic weapons. In particular, the United States wanted to consider its ICBMs and submarine-launched long-range missiles as strategic but not its 7000 or so nuclear weapons stationed in Europe and on the surface fleet, even though these are so deployed as to be able to demolish the U.S.S.R. The Soviets apparently wanted to consider all their missiles that are able to reach the United States, but not all those able to reach our European allies. Such an asymmetry can lead to a long procedural wrangle if there is at least one side not anxious to make an agreement.

Finally, in May, 1971, a joint communiqué was issued indicating not that any agreement had been negotiated but merely that there was agreement on what to start negotiating about. In view of the deadlock over defining strategic arms, the U.S.S.R. had wanted to negotiate an ABM limitation. The United States had apparently wanted to put numerical ceilings on other missiles while permitting further development of their quality. In the compromise announced, the United States gave in enough to agree to concentrate on ABM agreement while the Soviets agreed to negotiate on "certain measures" limiting offensive weapons.[19]

This came just at military budget time in Congress in May, 1971, when the opposition to the ABM and MIRV appropriations was particularly severe, and President Nixon advertised the communiqué with great fanfare, hailing it as a "major breakthrough" to a "new era" when "all nations will devote more of their energies and their resources, not to weapons of war, but to the works of peace." It was being made clear to Congress at the same time that he would need the ongoing programs of ABMs and MIRVs as bargaining chips.

Under Eisenhower, John Foster Dulles wanted to "negotiate from strength." Under Kennedy the phrase was "arm to disarm." President Nixon and his Secretary of Defense talked of "bargaining chips" for SALT. One can wonder how much of the incentive to carry on arms-control talks like SALT and disarmament talks like the Conference of the Committee on Disarmament arises from the political necessity to keep the doves content with military spending at home while seeming abroad to meet obligations like Article VI of the Nonproliferation Treaty. One can wonder whether the decision-makers see beyond the next budget, whether they appreciate the full horror of a perpetual arms race with its likely consequence, whether they perceive how little difference a few of the fancy multi-billion-dollar items in the budget make to the effectiveness of deterrence, which remains stable with only a very rough equality of overkill between the two sides.

The SALT One Agreement

The outcome of about three years of the SALT negotiations was announced in connection with a summit conference, which was held principally between President Nixon and Communist party chief Leonid Brezhnev in Moscow in May 1972. The agreement is sometimes called SALT One in anticipation of more to follow. It showed that the decision makers did indeed appreciate the dangers of an unlimited arms race and that both sides recognized that rough parity between the nuclear capabilities of the two powers was sufficient. The agreement was not a step to disarmament; it provided for no reductions, but it did put a definite numerical lid on the arms race. After so many years of blind racing it was a real achievement.

The numbers of weapons permitted were mostly somewhat above the current ones: just the current numbers of ICBMs and U.S. missile submarines, a substantial increase in MIRVs, with the Soviet Union permitted to catch up in submarines and multiple warheads, and a relatively modest limit on ABMs. The ABM agreement was by formal treaty, the rest by executive agreement for five years. The numbers agreed on are given in the following table.

The agreement does not refer explicitly to the fact that the United States has about 7000 nuclear missiles in Europe, many of them capable

TABLE 9–1
Numbers in SALT One agreement

	U.S.	U.S.S.R.
ABMs	200 (N.D. and maybe Washington)	200 (Moscow and one missile site)
ICBMs	1054 (constant since mid-60's)	1550 (with larger life capacity)
Submarines	41 (constant since mid-60's)	42 (finishing those being built)

of reaching well into the Soviet Union and many based on surface ships, but this fact is compensated in achieving rough parity by the larger number of ICBMs the U.S.S.R. is permitted to have, many of them being larger than those of the United States. (See Appendix 17.)

Although a definite lid was put on numbers, the SALT One agreement did not succeed in stopping the qualitative arms race. Both sides are permitted to continue weapons development and substitute new weapons for old ones within the numerical limits. One reason presumably is that compliance with a development ban would be more difficult to verify. Two verifiable limitations on development that should follow in subsequent agreements are a complete ban on nuclear tests, including underground tests, and a ban on missile test firings (perhaps with controlled exceptions for routine maintenance tests). Also, an agreement limiting "killer" submarines to smaller numbers than missile submarines would foster confidence in the undersea deterrent weapons.

The same pressures for large arms-race expenditures that have been effective for so long can continue and even press for increased vigilance under the imposed lid, similarly as increased testing, though underground, followed the partial test ban of 1963. Yet the effectiveness of this view should be dulled, both by official recognition of the sufficiency of rough parity, and by the lack of a large and uncertain opposing ABM system to be penetrated. With the sufficiency of the deterrent so amply assured, either side should feel confident in exercising some unilateral restraint in development with the expectation that the other would to some extent reciprocate. On each side, pressure should be felt to divert the funds and effort to more constructive ends.[4] The historic significance of the agreement will depend on which view prevails.

PROJECT PLOWSHARE

The people skilled in making nuclear bombs rather envied the reactor people the civilian industrial uses to which their technology was put and in about 1959 invented some nonmilitary uses for nuclear explosives. The program for promoting these was launched in the AEC as Project Plowshare, after the biblical phrase (from Isaiah 2:4) "They shall beat their swords into plowshares." The purpose of a plowshare is to disturb the earth and it is hoped to do this in a big way with nuclear explosives.

The two principal Plowshare objectives are

1. Nuclear blasting to make canals and harbors.

2. Recovery of natural resources: gas, oil, and metals.

The argument for these objectives is that they may save labor and thus money. Arguments against them are that they introduce serious radioactive contamination either of the region or the product, as well as of the broader environment, and that saving labor means creating unemployment in a society in need of finding an equitable basis for distributing goods and providing incentive for useful activity. The work that can in principle be done by Plowshare can be done in other ways.

The digging of a new Panama Canal is the prime example. By the rough method of nuclear blasting, one would blast a sea-level canal over a special route (one of several) at least a hundred miles from the present canal (and twice as long) to avoid destruction of the two cities at its ends. It would take about 200 MT, the largest explosives being 30 MT at the mountain stretch. It would cost about a billion dollars, about the same as another high-level canal alongside the present canal but with larger locks to take the increasing traffic. Nuclear blasting is said to be cheaper only because conventional digging *at the site chosen as best for the nuclear method* would cost about five times as much.[31,33]

But the nuclear method has political costs in addition to the engineering costs. It would be necessary to move about 30,000 rather primitive people away from their jungle for some years to avoid the radioactivity. This would be reminiscent, but in a peaceful operation, of the brash way our military relocates survivors in Vietnam. Further, it would probably be necessary to make arrangements for special exceptions to the partial test ban treaty to accommodate the radioactive winds from the larger cratering blasts.

This is typical. The underground blasts in gas-bearing rock to loosen the gas run into such radioactive trouble that a considerable part of the gas has to be vented with its fission-product burden into the atmosphere before the later and less radioactive gas is mixed with other gas to be piped to where it can spread the diluted radioactivity out among many users.[32] The partial test ban treaty permits underground tests on the presumption that they will not greatly contaminate the atmosphere, with the practical limit that the radioactivity shall not be detectable across national boundaries. Releasing into the atmosphere the first part of the gas produced, containing some of the fission products (plus tritium) and too radioactive to be diluted in the pipelines, seems to nullify part of the reason for permitting the underground test, for it deliberately vents a considerable part of the radioactivity into the air. The radioactivity can be released slowly, in contrast to the sudden burst from an above-ground test, and this makes it less easy to identify across national boundaries. Further, it is only a modest part of the totality of fission products produced by the nuclear explosive. The solid products remain embedded in the glassy rubble, and the venting can be delayed long enough to permit the short-lived part of the radioactivity in the gaseous fission products to have decayed. Thus the contamination of the atmosphere may be a less serious objection to this procedure than is the prospect of piping low-level radioactivity (even if well within official tolerance limits) into the kitchens and gas-heated homes of the nation.

Of the ABM, a weapons system promoted in the late 1960s, one month as anti-Soviet and another as anti-Chinese, it has been said that it was a "weapons system in search of a mission." The same might be said of Plowshare. The extent to which these essentially unneeded though potentially profitable techniques have been pushed is a tribute to the political power of their promoters. However, the canal project had its development funds cut for fiscal 1972 and seems to be politically dead.

From the point of view of diplomatic national aims, the promotion of Plowshare has backfired. The idea has been oversold abroad. The great technical difference between reactors and bombs has made it possible for reactors to spread to many countries without complicating the problem of proliferation of nuclear weapons. The problem is difficult but hopefully manageable through the Nonproliferation

Treaty (NPT) and further means of control to be developed. But Plowshare has made other countries want nuclear explosives as well, and
this introduced a real complication in pressing for the NPT. Brazil was
the leader of countries insisting that, while they could give up the right
to test nuclear weapons, they would not give up the right to test nuclear
explosives. Operationally, there is no distinction between such tests.

The difficulty has been surmounted by writing into the NPT a
provision that the nuclear-weapons nations will supply to the others
at cost (not including development costs) the service of carrying out
nuclear detonations for engineering purposes. During the course of
promoting the treaty with this provision, the AEC released information
about the cost of nuclear explosives that had hitherto been highly
secret, although it had been appreciated in a general way that nuclear
explosives in comparison with conventional explosives are an extremely
cheap means of destruction—one reason the nuclear age with its
prospect of proliferation is so dangerous. The price list released in
1968 by the AEC is:

Power of explosive	Cost (exclusive of development)
20 kT	$350,000
200 kT	500,000
1 MT	600,000

The megaton explosive is clearly the large economy size. The small
increase in cost with power—fifty times the power of the 20-kiloton
explosive for less than twice the price—seems to suggest that most of
the cost of an H-bomb is in the fission trigger. If the cost of development were included, the cost per bomb would doubtless be much higher
than this, particularly for a smaller nation stockpiling relatively few
of them.

While this arrangement in the treaty would really save the smaller
countries a great deal of development expense if Plowshare were
practical, there has been some reluctance to accept this dependence on
foreign sources for fear of some sort of commercial discrimination.
Thus the overselling of Plowshare has added to the difficulty of obtaining the Nonproliferation Treaty and will probably add to the difficulty
of making it effective in the future.

INDUSTRIAL RECONVERSION

President Eisenhower's phrase "unwarranted influence of the military-industrial complex" alludes to the strong influence wielded by industrial prosperity on decisions to go on developing and procuring ever more sophisticated arms. This influence comes not only from industrial or military management. It comes also from the workingman's level. Men want jobs, and they vote accordingly. Legislators who must be interested in reelection are of course sensitive to pressures the military can bring to bear through the allocation of military contracts in their districts. The problem of approaching disarmament then involves not only the design of a system of checks and balances between nations but also the internal political problems of finding other occupations for people engaged in making armaments, and of planning for economic reconversion of armament industries. In the long run, with the growing sophistication of techniques of producing goods based largely on the availability of energy sources, far-reaching modifications may be needed in our whole system of distributing the material needs of people while providing incentives for them to lead useful and rewarding lives. But in the short run, finding ways to retrain workers and retool industries to contribute to our many unmet civil needs is a part of the politics of nuclear disarmament.[4] This is one aspect of the social challenge of nuclear energy.

The biblical source of the name "Project Plowshare" is, more completely, "They shall beat their swords into plowshares, neither shall they learn war anymore." If Plowshare could have been phased in while disarmament was phasing out the development and deployment of military nuclear explosives, it would have been a good example of industrial reconversion, employing some of the same skills of workers (though probably too few workers to help much). Nuclear swords would indeed have been beaten into plowshares. Embarking on space exploration provided another such opportunity that was lost. As it is, "Plowshare" has gone hand in hand with learning war techniques of nuclear swords, the two contributing to each other in many series of underground tests.

THE SOCIAL CHALLENGE

Mankind has grown little if any in wisdom and stature since the days when he depended on his muscle for power and on a wood fire for his heat, but he has grown mightily in his command of nature. He has tasted the apple of the new nuclear firc and he both lives under its terrible threat and holds in his hand its hopeful promise. His wisdom is challenged as never before. He has age-old habits of fighting his fellow man for the scarce support of life, habits of harsh living that kept his numbers down. His new command of nature makes life less harsh and his numbers increase alarmingly. If he does not rise above his fighting habits, the nuclear fire could keep his numbers down in hideous ways from which he might never recover.

Over large areas the support of life is richly abundant. The promise of the nuclear fire can make it even more abundant there for a while or it can spread and prolong the present abundance. Yet the spread of the fire in its promise can also increase its threat. The balance is delicate, and man in his complex society must grow in his knowledge of the new fire and use his wisdom well if he is to reap its promise and evade its threat. This is the social challenge posed by the nuclear fire to his wisdom. To the increase of his knowledge of the fire and his awareness of the challenge, this book is dedicated.

REVIEW QUESTIONS AND PROBLEMS

1. Was the "Baruch plan" originated by Bernard Baruch? Did it propose complete world government? What was its relation to the Acheson-Lilienthal plan? Did it propose giving the "secret" of the A-bomb to the world with no concessions by others? What werc its provisions concerning uranium supplies?

2. What was the Soviet counter-proposal to the "Baruch plan"?

3. After Khrushchev came into power, there was a rather dramatic change in Soviet disarmament proposals, accepting many points proposed by the west. When was that, 1955, 1959, or 1963?

4. Did that proposal envisage giving access of international inspectors to entire land areas at only specified times or to all objects of control at all times? How did the United States react?

5. Were the first serious international discussions of a test ban treaty before or after the election of President Kennedy? What snag did they run into? Were they before or after the first nuclear test underground? Were they motivated only by the desire to get rid of radioactive fallout from the tests?

6. Was a limited test ban treaty, still permitting underground tests, first promoted by the Soviet Union or by the West? What year did such a treaty come into effect?

7. In a period of almost three years, 1958–1961, there was no nuclear testing by either side. Which side started the testing after that? Was any treaty or solemn obligation broken thereby?

8. What is meant by a low-level stabilized deterrent?

9. What specific fears have prevented nations from agreeing on general and complete disarmament?

10. How might it be possible in principle to bring about a balanced reduction of nuclear armaments to a low level without either side risking that inspection would uncover all its military secrets and open it to possible attack?

11. What danger might there be in having many more nuclear nations? About what fraction of the nations of the world are parties to the nonproliferation treaty?

12. In what way has the proposed engineering use of nuclear explosives hindered the attainment of a nonproliferation treaty?

The Language of
Large and Small Numbers

In these pages we have occasion to refer to many very large and very small numbers. It is worth taking the trouble to learn the special language for these numbers. They are often most conveniently expressed as powers of ten. A hundred is 10 times 10, that is, 10 squared or 10 to the power 2: $100 = 10 \times 10 = 10^2$. The exponent 2 indicates the number of times we have multiplied by 10. Similarly a billion is 10 times 10 times $10 \cdots$ increasing to nine factors of 10 altogether: $1,000,000,000 = 10 \times 10 \times 10 \cdots = 10^9$. The exponent also gives the number of zeros in the longer expression, each zero indicating multiplication by 10. We know in words that a billion is ten times a hundred million, but it is also easy in terms of exponents, $10^9 = 10 \times 10^8$, because we appreciate at a glance that 10^8 is one-tenth of 10^9—it lacks just one power of ten. Similarly a billion is a thousand times a million, $10^9 = 10^3 \times 10^6$, and we see how the exponents add to the same figures on the two sides, $9 = 3 + 6$.

Small numbers can be expressed with exponents, too, but then the exponents are negative. One-tenth is 10 to the minus-one power: $\frac{1}{10} = 10^{-1}$. This way, the exponents still add up nicely. For example, $10^8 = 10^{-1} \times 10^9$ says that a hundred million is one-tenth of a billion, and the exponents on the two sides add to the same quantity. Now it is much easier to write and think 10^{-8} than "one one-hundred-millionth." The negative exponent tells how many times we have divided by ten.

It may take a little practice to get accustomed to this usage. By way of practice and review, consider the first page of the Introduction rewritten in terms of the powers of ten.

Age of known universe:	5×10^9 years
Carboniferous era:	5×10^8 years ago
Duration of that era:	10^8 years
Man evolved:	10^6 years ago
Man's use of fire:	10^4 years
Wind and water mills:	10^3 years
General use of coal:	200 or 2×10^2 years
Petroleum:	100 or 10^2 years

Another language for large and small numbers employs prefixes based on Greek and Latin roots thus:

10^{12}		trillion (in U.S.)
10^9	giga	billion (in U.S.)
10^6	mega	million
10^3	kilo	thousand
10^{-3}	milli	thousandth
10^{-6}	micro	millionth

Kinetic Energy

Suppose that a body starts from rest, so that it has no kinetic energy, and is accelerated with a constant acceleration a for a time t. Its final velocity at the end of the time is

$$v = at \quad \text{so that} \quad a = v/t.$$

Constant acceleration is the result of a constant applied force $F = ma$, and the work done in accelerating it through a distance d to a final velocity v is

$$\text{Work} = Fd = mad.$$

The distance traveled while the speed was increasing uniformly from zero to v is

$$d = v_{\text{average}} \, t = \tfrac{1}{2} \, vt$$

since the average velocity (the average of zero and v) is $\tfrac{1}{2}v$. Thus the final kinetic energy, or the work done in speeding up the body from rest, is

$$KE = mad$$
$$= m\frac{v}{t} \, \tfrac{1}{2}vt = \tfrac{1}{2}mv^2.$$

The quantity mv^2 can be expressed as a number of kg meter^2/sec^2, for example, which is consistent with (force) × (distance) or (kg meter/sec^2) × meter.

We see that the factor 1/2 comes in because it is average velocity that determines the distance. This may be emphasized graphically by drawing the diagram of Fig. A2–1. The diagonal line shows that

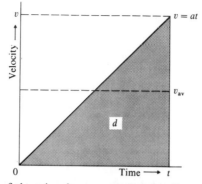

Fig. A2–1 The area of the triangle, representing the distance traveled, is half the area of the square.

velocity increases with time, $v = at$. This may be compared with Fig. 0–2, in which the curve gives the change per year of the energy used, and the area under the curve gives the total energy used. Here in Fig. A2–1 the line $v = at$ gives the change per second of the distance traveled, and the area under it gives the total distance traveled d. Thus the area of the shaded triangle is d. It is 1/2 the area of the square, at^2, or equal to the area of the lower half of the square, that is $v_{av}t$.

APPENDIX 3   Sound Waves
and Shock Waves

We are all familiar with waves on water. The wind blows and covers a lake with waves propagating at right angles to the straight-line wave crests. A pebble is dropped into a calm pond and a sequence of little waves radiates from the spot in concentric rings. Observation from a boat of bits of flotsam on a rough lake shows that the water on the surface moves forward on the crest of the wave and at the same time backward in the trough of the wave. Think now of the front slope of the wave, the part between a crest and the next trough ahead of it. Water is converging on it from both directions. The surface there is being squeezed upward, so that soon it will be the highest point of the wave as the crest moves forward. This is part of the mechanism that keeps the wave traveling while the water itself just moves back and forth making no net progress, staying approximately where it is.

The mechanism of a sound wave in a gas like air is very much like this. In place of crests and troughs on a surface, a sound wave in air is made up of regions of high pressure and low pressure. Air in the high-pressure region is moving forward and that in the low-pressure region moving backward, with appropriate accelerations in the regions in between. Air just ahead of the high-pressure region is being squeezed by the influx from ahead and behind, to make it the future high-pressure region. In fact, the cross-sectional picture of the water wave in Fig. A3–1(a) may be viewed as a graph of the pressure in the air wave. The arrows indicating local motion could also represent the exactly analogous situation in the sound wave.

The speed of sound in air depends slightly on the average pressure of the air. Sound moves slightly faster through higher-pressure air.

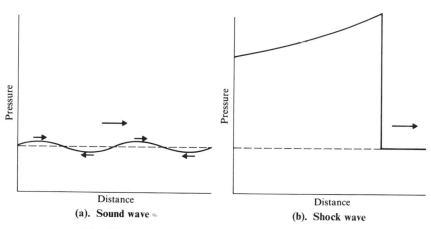

Fig. A3–1 Pressure profile of sound and shock waves.

Ordinary sound is a long train of similar waves. (If the distance from crest to crest, or wavelength, is relatively short, the frequency with which waves arrive at the ear is high and the sound has a high pitch.) A shock wave is quite different. It is a single pressure wave with an extremely sharp rise of pressure at its leading edge. The back of the shock wave is a gradual diminution of pressure. The sudden change of pressure at the leading edge is so great that the speed of sound (and other pressure disturbances) behind the shock front is appreciably faster than in the undisturbed air in front of it. This means that the back of the shock wave and any small pressure fluctuations in it, are continually trying to catch up to the shock front and add to the energy it represents, without being able to go on ahead of it. Looked at from behind, the rapidly moving shock front is where the higher speed stops.

Since a shock wave depends on a big enough change of pressure to increase the speed of sound appreciably, it occurs only when a lot of energy is suddenly put into a strong compression. This happens in the air around various types of explosions and is the principal way that energy is carried away from some explosions. The shock wave spreads as an expanding sphere about a point of explosion, and its intensity falls off as the inverse two-thirds power of the distance. This means that the blast from a big nuclear bomb is one-fourth as strong at eight miles as it is at one mile, for example.

Shock waves can also exist in liquid and solid substances. In a high-explosive solid, such as TNT, the shock wave can actually build up in intensity as it progresses. The sudden rise of pressure detonates the explosive as it passes. The energy released creates an additional increase of pressure which quickly catches up to the shock front. By shaping the surface at which it is initiated, a shock front in an explosive can be focused on a point or line to make extremely high pressure for an instant. Such shaped charges are used to puncture holes in thick armor plate.

APPENDIX 4 Mechanics of Atoms

Atoms are so much like the solar system that we shall try to understand both in the same language. It is perhaps easier to think about the solar system because we can see some of the planets in the sky and cannot see the individual electrons in the atom. The remarkable thing about a planet orbiting about the sun or a comet approaching the sun is that it is continually attracted toward the sun but never falls into it. The situation of electrons in atoms is similar to this. To understand it we need to know more about acceleration.

ACCELERATION AND FORCE IN STEADY CIRCULAR MOTION
Until now we have discussed mainly that aspect of acceleration that consists of speeding up or slowing down motion along a straight line, but we have mentioned that there can also be a sideways aspect to acceleration. This is important for understanding orbits.

If a car going north at a steady speed veers slightly—only a degree or so—toward the west, the sideways force and sideways change of velocity and momentum are essentially westward. The rate of change of the velocity vector (the acceleration) is westward as the velocity vector veers from the north toward the west. When a vector has constant length but changes in direction, its rate of change is at right angles to the vector itself.

The idea that the turning of the velocity vector corresponds to a sideways acceleration may seem a little abstract. Perhaps some further discussion in terms of the acceleration of a stone in a slingshot may help to clarify the idea. If you are standing still when you shoot the stone, the slingshot speeds up the stone along a straight line; it causes a longitudinal acceleration. But imagine that you are in an open automobile traveling fast northward along a straight road and shoot the slingshot directly sideways, toward the west. As you grasp the stone, it has a high velocity northward. But just as you let it fly, while it is still in the slingshot and just starting to move away from your fingers, it has an acceleration toward the west. The westward part of its motion starts off slowly (from zero) and speeds up. If you could trace the actual path of the stone as a person on the ground would see it, it would follow a curved line that starts off northward but curves over toward the west. In fact, it would be following quite closely a small arc of a circle. This gives a feeling for how motion around a circular path is associated with an acceleration at right angles to the velocity. It is just as truly an acceleration of the stone in the slingshot, but the motion is longitudinal in one case and turned sideways in the other.

The same kind of motion around a curve can be made by attaching the stone to a string and swinging it around in a circle. The string exerts a force toward the center that makes an acceleration toward the center. The velocity is at right angles to the string, and the acceleration is at right angles to the velocity, which means along the string. At the instant the stone is moving north, the string may be accelerating it toward the west, just as in the case of the slingshot in the automobile. Let the speed of the stone be constant as it steadily travels around in a circular path. The center of the circle may be considered the origin or reference point, so the string itself, with an arrow pointing outward, is the radius or position vector. In a short time the stone moves from position A at the tip of vector r_A to position B (with radius vector r_B) in Fig. A4-1. The distance from A to B is equal to the velocity times the time. A and B are assumed to be close enough together that the velocity does not change much during that short time. It does not change in magnitude—it may remain 2 cm/sec, for example, throughout the time—but it does change a little in direction, for it remains always at right angles to the string as the string rotates.

At the right side of Fig. A4-1 there are two velocity vectors, the velocity when the stone is at A and the velocity when it is at B, and the short arrow between the points of the two vectors is the amount by which the velocity has changed. Acceleration tells how rapidly velocity is changing with time, just as velocity tells how rapidly position is changing with time, so the length of this short arrow is (acceleration) × (time). Therefore we have two narrow triangles in Fig. A4-1. They

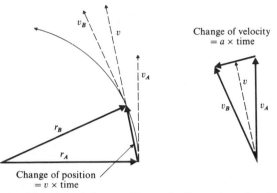

Fig. A4–1 Vector triangles relating position, velocity, and acceleration in steady motion around a circle.

are of different sizes but the same shape because there has been the same amount of rotation between the long sides in either case. (In geometry we say that they are similar triangles.) One triangle is simply a scaled-down version of the other (and drawn at a different angle on the page). This means that the ratio formed by dividing the length of the short side by the length of the long side has the same value for both:

$$\frac{\text{short side of one}}{\text{long side of one}} = \frac{\text{short side of other}}{\text{long side of other}}.$$

Inserting the actual lengths of these sides, we have instead the expression

$$\frac{(\text{acceleration}) \times (\text{time})}{(\text{velocity})} = \frac{(\text{velocity}) \times (\text{time})}{(\text{radius})}.$$

If we remove the (time) factor from both sides and multiply both by (velocity), we have

$$(\text{acceleration}) = \frac{(\text{velocity})^2}{(\text{radius})}.$$

Here the expression (velocity)² means (velocity) × (velocity). As an example, the velocity might be 5 cm/sec and the radius might be 10 cm. In that case the acceleration would be

$$(\text{acceleration}) = \frac{(5 \text{ cm/sec})^2}{10 \text{ cm}} = 2.5 \text{ cm/sec}^2.$$

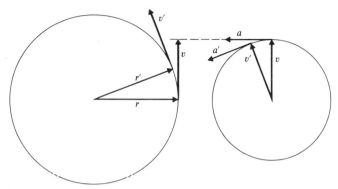

Fig. A4–2 Motion of the position and velocity vectors around circles.

Let us derive again this same acceleration characteristic of steady circular motion, using circles instead of the triangles.

The position of the stone swinging on a string around a circle may be described by a radius vector pointing outward along the string. As shown in the first derivation, this is a position vector and rotates at a steady rate. The instantaneous motion is at right angles to the string (since the string does not stretch and permits no motion along its direction). If the string should break and exert no more force, the body would "fly off" by continuing along a line tangent to the circle. This is represented at the left side of Fig. A4–2, which is drawn to scale. In it, the vector r represents a length: a centimeter might represent a meter, for example. But the tangent vector v represents a number of meters per second, the rate of change of the vector r as it rotates around the circle. The velocity is shown as v' at the instant when the object passes by the point indicated by vector r'. In the diagram on the right side of Fig. A4–2, the various vectors v and v' are drawn starting from the center, as r and r' were in the other diagram. The length of v represents a number of centimeters per second, cm/sec. The tangent vector a represents a number of (centimeters per second) per second, cm/sec^2, the rate of change of the vector v as it rotates around the circle. The velocity vector on the second diagram is at each instant at right angles to the radius vector in the first diagram and thus rotates at the same steady rate. Its length is constant because the rate of rotation is steady. So, while the point of the radius vector goes around one circle describing the path or orbit of the body, the point of the velocity vector goes around the other circle, but always ahead by 90°. The time taken to go around each circle is the same. For a given rate of rotation, the rate at which a point progresses around a circle is of course proportional to the radius of the circle: the bigger the circle, the faster the point has to go to get around in the same time. The position vector moves around its circle (of radius r) with the speed v and the velocity vector moves around its circle (of radius v) at a rate a, the magnitude of the acceleration. Since these rates of moving around the circles are proportional to the respective radii, we have

$$a/v = v/r \quad \text{or} \quad a = v^2/r.$$

This is an acceleration that depends only on the motion itself, only on the fact that the object is moving on a curved circular path at a given speed, and does not require thinking about the force that makes it move that way.

However, when a body is accelerated we have learned that there is a force acting on it (in this case the force exerted by the string) given by

$$\text{force} = (\text{mass}) \times (\text{acceleration}).$$

Thus we may write a word equation

$$\begin{pmatrix} \text{force toward the center required to} \\ \text{keep a mass moving steadily in a circle} \end{pmatrix} = \frac{(\text{mass}) \times (\text{velocity})^2}{(\text{radius})}$$

or, in shorter symbols,

$$F_c = \frac{mv^2}{r}.$$

We call this F_c, rather than just F, to remind us that it is the force toward the center characteristic of steady *circular* motion. It is sometimes called centripetal force which means "center-seeking" force.

ORBITS IN ASTRONOMY

Orbits occur in many different sizes. There are the large astronomical orbits of planets and comets about the sun and the orbits of satellites about the planets. There are the tiny orbits of electrons in atoms and the even very much smaller orbits of nucleons in the nucleus. These "orbits" are somewhat indefinite because of their small size, and only some of the features of larger orbits apply. While our main interest will be in the small ones, it is conceptually easier to start with the large orbits, the evidence of which we may see in the sky with our eyes and with telescopes. We see, for example, the changing phases of the moon as evidence that it circles the earth in about 28 days.

The Inverse-Square Force of Gravity

The surface of the earth is at a distance $r = 6000$ kilometers from its center. Imagine dropping an object at rest (that is, not rotating about the earth) from a position twice as far from the center of the earth, at a radial distance $r = 2r$. (To get it there, we might imagine building a tower 6000 kilometers high at the north pole!) What would happen? The answer is that the object would drop with an acceleration $a = \frac{1}{4}g$, where the factor $\frac{1}{4}$ is $1/2^2$. That is, the acceleration is inversely proportional to the *square* of the radius, or a is proportional to $1/r^2$. This acceleration of the falling stone is caused by a force acting on it, given by

$$\text{force} = (\text{mass}) \times (\text{acceleration}),$$

so also the force exerted on the stone by the gravitational pull of the earth is inversely proportional to the square of the radius. It is also proportional to the mass of the stone. As can be seen by watching astronauts hop around on the moon, the acceleration due to gravity on the moon is less than it is on the earth because the moon has a smaller mass than the earth. Our falling stone would accelerate more slowly on the moon. The force of gravity on it is proportional not only to its own mass but also to the mass of the body attracting it. It is a fundamental law of nature that any two material bodies, say with masses M and m, attract each other with a gravitational force, here called F_g, that varies with the masses and the distance r between their centers thus:

$$F_g = \frac{GMm}{r^2}.$$

This G is a constant that is always the same. If we should want to, given the mass of the earth, we could figure the value of G by relating it to the sample of gravitational acceleration we observe at the surface of the earth:

$$\frac{G \times (\text{mass of earth})}{(\text{radius of earth})^2} = 9.8 \text{ meters/second}^2$$

which is called g. Thus at the surface of the earth $F_g = mg$. There is a gravitational force between two stones held in your two hands but it is so small that you cannot notice it. This is because G is so small that at least one of the masses has to be huge—like that of the earth—before the force of gravity F_g is ordinarily noticeable.

How can we show that F_g has something to do with everyday experience, without building that tall tower? Let us begin by showing that it helps us to understand the relation between the fact that it takes the moon about 28 days to circle the earth, as we know from watching it in the heavens, and that it takes an astronaut about an hour and a half to do the same thing just above the earth's atmosphere, as we know from reading the papers. The moon and the astronaut are each moving in a roughly circular orbit about the center of the earth but with different values of speed v and radius r. For a given r there is just one value of v with which an object can move in a circular orbit. This is obtained by requiring that the gravitational force be just the center-seeking force required to maintain the circular orbit, that is, that $F_g = F_c$. This velocity is such that the time it takes the orbiting object to go once around the circular path (the period of the orbit) is proportional to $r\sqrt{r}$, that is

$$t = \text{const} \times r\sqrt{r}$$

as is shown with a few lines of algebra in Appendix 5. The main point to be drawn from this equation is that for a larger orbit the object both has farther to go and also moves more slowly to match the smaller gravitational force, so naturally takes more time to get all the way around.

Now the radius of the moon's orbit is about 60 times the radius of the earth R, so for the moon $r = 60R$, and the time for the astronaut's orbit approximately at radius R may be written

$$(\text{time for astronaut's orbit}) = \frac{R\sqrt{R}}{60R\sqrt{60R}} (\text{time for moon's orbit})$$

$$= \tfrac{1}{465} \; 28 \text{ days} = 1\tfrac{1}{2} \text{ hours}.$$

This is about the time it actually takes. A small correction can be made because of the fact that artificial satellites usually have an orbital radius of about $1.05R$ so as to clear the earth's upper atmosphere. This discussion of orbiting time is of course a digression that does not directly help us understand nuclear energy. It merely illustrates how nicely our ideas about force and acceleration work and why scientists have confidence in them while applying them to the understanding of atoms and nuclei.

The condition of having the gravitational force remain exactly equal to the force F_c required for a circular motion may sound like an unstable situation, something like trying to roll a bowling ball down the middle line of a bowling alley that is warped downward toward both sides.   If the moon were to be deflected slightly outward from its circular path, would it fly off to infinity? If it were deflected inward, would it spiral inward to hit the earth? The answer is no, the motion in the orbit is quite stable and these small deviations tend to be corrected. It is more like rolling a ball along the bottom of a valley.   Any small deviations are automatically corrected.

To get some feeling for orbital stability, consider a more general type of motion.   The circular motion is a special case, the simplest one to describe, but a planet or satellite can also move in an elliptical orbit, with the sun, in the case of a planetary motion, at what is called a focus of the ellipse.   This is a point inside the ellipse near one end when it is a long ellipse as in Fig. A4–3   Actually most comets move in such long elliptical orbits, and are nearest the sun when at one end of the ellipse. The earth and other planets move in nearly circular orbits that are actually fat ellipses.

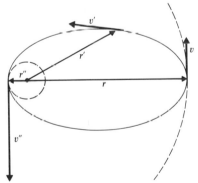

Fig. A4–3 Change of position and velocity around elliptical orbit.

While thinking of orbits of comets and electrons, it may be helpful to think at the same time of a similar motion on a scale that is more familiar.   As an example of an orbital motion under the influence of something like an inverse-square force toward a center, we may think of sliding a cube of ice or dry ice with no friction in an appropriately shaped large circular dish, one that is nearly flat near the edge and gets

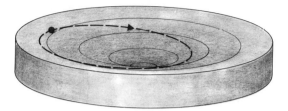

Fig. A4–4 A dish that gets steeper near the center, in which a ball may roll in an approximately elliptical orbit.

progressively steeper toward the center, as sketched in Fig. A4–4. (It should have an inside profile about like the curve shown in Fig. A4–6, page 270, but perhaps flatter.)

When a comet is approaching the sun, as at the point r' in Fig. A4–3, the force of gravity is pulling on it harder and harder as it gets closer, according to the inverse-square law. One might wonder why, in the clutches of such an increasing force, it does not fall into the sun, spiraling ever inward. The reason that it does not also helps to explain why the solar system and atoms continue to occupy the space they do. A rough understanding may be had by thinking of an automobile gaining speed as it rolls downhill and then having to turn a sharp corner at the bottom. The auto is gaining speed because it is getting nearer to the center of the earth (losing *PE* while it gains *KE*). The comet is doing the same thing as it approaches the sun. If at the bottom of the hill the road has a sharp curve that should be driven at 20 miles per hour but the auto has speeded up to 60 mph, it will fail to round the curve in spite of having good tires that provide a lot of friction. That is similar to the situation with the comet; when it gets in close it has to turn very sharply to stay close, and it is going so fast that it doesn't quite succeed in spite of the strong inward pull of gravity.

To understand the elliptical motion a little better, consider first a simpler example of a ball tethered to a fixed center by a stretched spring. A spring is quite different from a string, for it may be stretched a lot and the farther it is stretched, the larger the force it exerts on whatever is stretching it. When it is at radius r, the ball may be given a velocity v at right angles to the radius and we may ask, "Will the ball travel in a circle?" The answer is that it will travel in a circle (of radius r) only if the spring is stretched just enough to provide the centripetal force

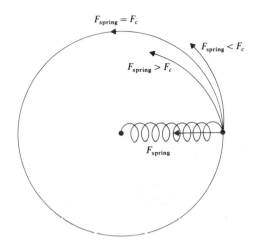

Fig. A4–5 Potential-energy curve corresponding to an inverse-square attractive force, indicating the range of variation of potential and kinetic energy in a typical elliptical orbit.

characteristic of uniform circular motion, in other words only if

$$F_{\text{spring}} = F_c = \frac{mv^2}{r}.$$

If instead the force provided by the spring is greater, $F_{\text{spring}} > F_c$, then the ball will follow a path that is curved more sharply than the circle of radius r at that point and will fall inside the circle, as in Fig. A4–5. If $F_{\text{spring}} < F_c$, on the other hand, the spring is not pulling hard enough to hold the ball in the circular path, and it will initially follow a path that goes outside the circle. Whether it will later come back to the circle is another question.

Consider now a comet (or the ice cube in the dish) moving along the outermost part of its orbit, at the right hand side of Fig. A4–3. The dotted line shows a circle tangent to the ellipse at that outermost point. There gravity is too strong to permit the comet to move along the circular path. That is, the velocity v that appears in the expression mv^2/r for F_c is small enough to make the force required for steady circular motion less than the inverse-square force of gravity. The comet thus falls inward along the elliptical path bringing it closer to the sun, pulled in by the gravitational force. When the comet is at the intermediate position r' in Fig. A4–3, it is moving partly in the direction of the applied force, inward along r', and gravity is doing work on it, increasing its kinetic energy as its potential energy decreases, just as gravity does work on an object falling from the table toward the floor. Its velocity increases so much that by the time it is in the innermost position r'' the force F_c that would be required to keep it in a small circle there (which is proportional to v^2) has become greater than the gravitational acceleration in spite of the fact that this, too, has increased. Thus, with $F_g < F_c$, the comet flies outside the small circle drawn with a dotted line at that point, outward along the outward leg of the ellipse, losing kinetic energy as it goes (because gravity is now pulling partly backward on it) until it arrives back at the starting point with its original kinetic energy. At this outermost point, where it started, its kinetic energy is minimum and its potential energy maximum, whereas at the innermost point, indicated by r'', the opposite is the case. The total energy remains constant all the while.

In Appendix 6, the situation is discussed in more detail. It is there shown that F_c does indeed increase so fast as to become greater than the gravitational force F_g as the radius becomes smaller. In that explanation the idea of angular momentum is used.

Angular Momentum

The angular momentum of a body moving about some center is defined as its radial distance from the center multiplied by its mass times the component of its velocity that is at right angles to the radius. A rotating flywheel has lots of angular momentum—a tendency to keep on coasting round and round just as a body with ordinary momentum has a tendency to keep on coasting along a straight line. In motions like the comet's, the angular momentum remains constant. This fact is helpful because it tells us how much the perpendicular part of the velocity, at right angles to the radius, increases as the radius decreases. It thus contributes to the idea that the comet (or electron) zips around the inner end of the ellipse too fast to stay in there.

For circular motions the angular momentum is simply the radius multiplied by the linear momentum. It remains constant as long as there is no cranking-type force applied, that is as long as the only force is along the radius. A laboratory example is a mass sliding on a table without friction and kept in a circular motion by a string that passes down through a small hole in the table at the center. If the string is pulled down to reduce the radius of the circular motion, the linear momentum and velocity are greater at the smaller radius. (Figure skaters and fancy divers know about this as they bunch up tighter to spin faster.)

POTENTIAL ENERGY WITH FORCE TOWARD A CENTER

The concept of potential energy is important because energy is a quantity that is conserved as it changes from one form to another. This is true no matter where we define the potential energy to be zero. The choice is arbitrary. For motions in a room the zero of *PE* may just as well be at the table-top level as at the floor. Below the table top the *PE* would then be negative. If *PE* is positive, it is a measure of the work that can be done by the gravitational force on the body as it goes down to the zero level. Below the table top the *PE* is negative and its magnitude is a measure of how much work has to be done against gravity to lift the body up to the position of zero *PE*. Doing work against gravity is the opposite of letting gravity do work.

When a particle is attracted toward the center with an inverse-square force, or with any force that is very small when r is very great, as in atoms and nuclei, it is convenient to define the potential energy as 0 when r is infinite, that is $PE = 0$ when $r \rightarrow \infty$. The force remains small where r is large, and this small force gives rise to small differences of potential energy. Therefore the potential energy curve is very flat for large r, becoming steeper for small r and curving downward to indicate that potential energy is negative, as shown in the curve of Fig. A4–6.

The ice cube sliding in the dish of Fig. A4–4 is a good example of motion with such a *PE* curve, for the shape of the dish can be made to reproduce the *PE* curve. If the cube is started with some angular momentum but not too much, near the outer edge it should follow an approximately elliptical path (if the dish is shaped to make an inverse-square force), as suggested by the dotted line in the figure. If the cube has just the right velocity it should achieve a circular path. In either case, the cube's total energy is negative; it would have to be started faster, with more *KE* and indeed positive total energy, to climb up the hill and over the edge of the dish. Its *PE* is determined by its depth below the zero level at what would be the height of the edge if the dish were very large ($PE = -mgd$, where α is that depth), just as given by the *PE* curve of Fig. A4–6.

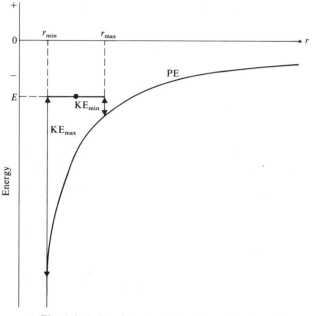

Fig. A4–6 Possible deviations from circular path.

As the cube slides in a roughly elliptical motion, traveling back and forth between a minimum radius and a maximum radius as in Fig. A4–3, its energy is continually being transferred from *KE* to *PE* and back again, almost as in the swinging of a clock pendulum except that neither *PE* nor *KE* goes to zero. At the outer radius the *PE* is at a maximum and the *KE* minimum, while at the inner radius the reverse is true. These characteristics of the motion may be described by the heavy horizontal line in Fig. A4–6, which indicates that the total energy *E* is constant as the radial part of the motion extends from r_{min} to r_{max}. The *KE* is everywhere the vertical distance between this level and the *PE* curve, being a maximum at r_{min} as indicated. A circular orbit would be represented by a dot at the appropriate *E* and *r*.

If the cube is started at just the right speed to move about in a circle. its motion is a model of the earth's orbit. Actually the orbit of the earth about the sun is not quite circular. It is an ellipse that is nearly circular and there is a small amount of this exchange of *PE* and *KE*, this speeding up and slowing down to make the orbit stable about its nearly circular shape. Thus the earth neither falls into the sun nor flies off into remote space. This situation is typical of all of the bound systems of particles that we deal with. Just as in an astronomical orbit, so the electron in its tiny orbit in an atom or the nucleon in its rather different orbit in a nucleus experiences a continual interchange between KE and PE. The PE tends to pull the particles together and the behavior of the KE tends to keep them apart enough so that the system retains its size and does not collapse to practically zero volume. Yet the KE does not influence the particles so strongly as to make them fly apart. Indeed, the atomic and nuclear systems that have survived in nature so that we may use them in our technology are those in which the tendency of the PE to hold things together is strong enough to do so in spite of the tendency of the KE to make them fly apart. The technology of releasing nuclear energy then consists of arranging circumstances in which some parts of nuclei do nevertheless fly apart.

ATOMS

One reason for the similarity between atoms and the solar system is that electric forces, like gravitational forces, are inverse-square forces. They are very different in that electric forces may either attract or repel, while gravitational forces always attract. In place of the gravitational force of attraction

$$F_g = \frac{GMm}{r^2}$$

between masses M and m that are always positive, we can describe the electric inverse-square force between two charges Q_1 and Q_2 as

$$F = \frac{Q_1 Q_2}{r^2}$$

where Q_1 and Q_2 can be either positive (as on the glass rod) or negative (as on the rubber rod). As we have seen, if Q_1 and Q_2 have the san. sign, F is positive, which means a repulsion. If one is positive and the other negative, F is negative and the two bodies will attract one another. (A negative repulsion is an attraction.)

A normal atom contains both positive and negative electric charge in equal amounts. The negative charge consists of electrons. We have already met the electron, the first of the subatomic particles. All electrons are exactly alike, and each carries a definite very small quantity of electric charge, the *magnitude* of which we call e. The charge e is a small number times a unit that we shall not here define. The charge on the electron is negative (like that on the rubber rod) so the charge itself is $-e$. Two electrons a centimeter apart, for example, repel each other with a very small force indeed, since atoms are extremely small. Even at such a small distance, the force between two electrons is very small by macroscopic standards. The small but definite charge e is the smallest electric charge that is known to occur in nature.

Another important constituent of the atom is the proton. It has a positive charge of $+e$, exactly equal and opposite to that of the electron. Thus, in view of its sign, e might more properly be called the proton charge. The proton is much heavier, having a mass $m_p = 1840\ m$, almost two thousand times m, the mass of the electron (but still a very small mass by macroscopic standards, a bit over 10^{-24} grams).

The third constituent of the atom is the neutron. It is like a proton, with very nearly the same mass, but no electric charge at all. It is called a neutron because it is electrically neutral; it neither attracts nor repels other particles electrically. Protons and neutrons are called nucleons. A proton is a charged nucleon and a neutron is a neutral nucleon.

Atoms are extremely small, something like 10^{-8} cm in radius (that is, one hundred-millionth of a centimeter). But a tiny atom has a considerable resemblance to the big solar system, to the sun with all its planets in orbits at various radii (Figs. A4–7 and A4–8). The positive charge is concentrated at the center, in a nucleus of the atom that contains almost all the mass (even as the sun does in the solar system). In the place of the planets there are electrons, all identical to each other and each very light and negatively charged. Between the positive charge at the center and each of the negative electrons there is a force of attraction, and this is what holds the atom together. The like-charged electrons repel each other, but not strongly enough to disrupt the atom. The heavy nucleus at the center of the atom is extremely small indeed, about 10^{-4} as big as the atom. It is made up of protons and neutrons holding each other together by very short-range forces that are neither electrical nor gravitational in nature—and we shall discuss these special nuclear forces later. The numbers of protons and neutrons in a particu-

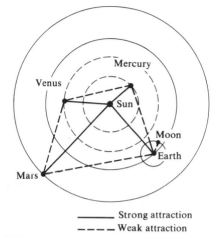

_____ Strong attraction
- - - - Weak attraction

Fig. A4–7 Strong and weak attractive forces in the solar system.

lar nucleus we denote by the symbols Z and N: a nucleus contains Z protons and N neutrons, or a total of $A = (N + Z)$ nucleons. Since only the protons are charged, the charge of the nucleus is Ze.

A normal atom also contains exactly Z electrons, and thus contains equal amounts of positive and negative charge, $+Ze$ due to the nucleus

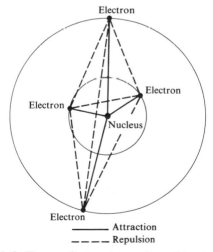

_____ Attraction
- - - - Repulsion

Fig. A4–8 Forces of attraction and repulsion in an atom.

and $-Ze$ due to the Z electrons each with charge $-e$. Though the electrons occupy a space that is somewhat fuzzy at the edges, the atom is nearly round like a ball. We may ask how much force the atom would exert on a charge Q_2 at a distance r, at a point far enough away that there the atom appears to be spherical. The answer is given by the formula we saw at the beginning of this section, $Q_1 Q_2 / r^2$ with $Q_1 = 0$. The atom has no charge, so there is no force exerted.

Let us consider this in more detail. If the charge outside the atom is a stray electron at a fairly large radius r from the nucleus, so that $Q_2 = -e$, the nucleus attracts it with a force $-Ze^2/r^2$ (along the direction of r, calculated with $Q_1 = Ze$) and the Z electrons each repel it with a force approximately e^2/r^2 (calculated with $Q_1 = -e$ for each of them), so there is as much repulsion as attraction, one strong attraction and Z weaker repulsions, as in Fig. A4–9. The net total is zero. There are electrons on both the far and near side of the nucleus and r is a bit short for some of them and a bit long for others, compared to r for the nucleus, so this evens out, and the variations in r can be ignored.

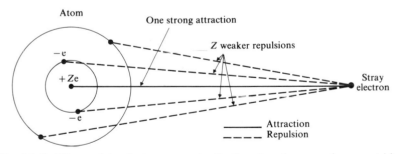

Fig. A4–9 The repulsion just compensates for the attraction on a charge outside a neutral atom.

All the electrons in an atom repel each other. For each electron there are only the $(Z - 1)$ other electrons repelling it while there are Z protons in the nucleus attracting it. Thus the attraction is stronger and the electron is held in the atom. An atom lacking one or more electrons is called an ion—a positive ion because it has a net positive charge. If there are any stray electrons around, it attracts them (for Q_1 is not now zero) and thus tends to make itself into a neutral atom of the kind found commonly in nature. Thus in a world made of nuclei and electrons, each nucleus attracts as many electrons as it can until its atom is neutral and it can't easily attract any more.

The following table summarizes the charges and masses:

TABLE A4–1
Particle charges and masses

Particle	Charge	Mass	
Electron	$-e$	m	
Proton	e	$m_p = 1840m$	
Neutron	0	$m_n = m_p$	
Nucleus	Ze	Am_p	$A = Z + N$

The simplest neutral atom of all is the hydrogen atom. Its nucleus consists of a single proton and it has only one electron. A positive ion with nuclear charge Ze but with only one electron is just as simple to discuss as the hydrogen atom, which has $Z = 1$. In discussing orbits of the earth, we could reasonably talk only about the pull of the sun on the earth, because the pull on the earth by the other planets is very much smaller. In most atoms this is not the case; the electrons interact not so very much more strongly with the nucleus than with each other, because electric charges, not masses, are making the forces. But the hydrogen atom is simple because there is only one pull on the electron. Here again we have a case of a light object moving under the influence of inverse-square attraction toward a heavy object, and for this an elliptical orbit would be expected.

Quantization

However, there is an important difference between the way we can describe motions in the submicroscopic realm of the atom and in the macroscopic realm of the laboratory or solar system. This difference applies to all of the electrons in a complicated atom or to the nucleons within a nucleus, but is most easily discussed in the simple hydrogen atom. Two important and related ideas are involved:

First, there is an intrinsic limitation of our ability to observe extremely accurately a position and motion of a particle confined within an extremely small space. This is known as the "uncertainty principle."

Second, there is an important natural constant, called Planck's constant and denoted by the symbol h, in terms of which the momentum or angular momentum of a particle can be described when the particle is confined to a very small space. We have seen that an object that feels a pull in along the radius toward a single center of attraction, such as a planet or a comet toward the sun, has a constant angular momentum. In an atom this constant can have only certain values, expressed in terms of h.

To illustrate how observation of position is limited, let us think of an electron in a very small and thin tube, bouncing back and forth between the two ends with a certain amount of kinetic energy. If it were a rubber ball in a big tube, one could describe where it was at any time. In the case of the electron, one cannot. Instead, one can only say where it might be, practically anywhere along the length of the tube. But we can say something about its momentum as it bounces back and forth. Its motion is "quantized." The quantum rule is that the magnitude of its momentum multiplied by the distance the electron goes on a round trip (in this case down the tube and back) must be an integer multiplied by Planck's constant. In symbols,

$$(\text{momentum}) \times 2L = nh$$

where L is the length of the tube and n is 1 or 2 or 3 or any such integer. This rule of the submicroscopic world can be discussed more fully in terms of a mysterious wave that guides the motion of these tiny particles (see Appendix 8), but for our purposes the rule itself will suffice. Planck's constant h is a very small but definite quantity. (It is actually 6.62×10^{-34} kilogram meter^2/second, which we see is the right kind of quantity to be a momentum, or mass times meters/second, times a length.) The reason that the quantum rule does not seem to apply to large objects like billiard balls or the moon is that the integer n would be so large that it could not be recognized as an integer; there is no practical difference between a trillion and a trillion and one.

We remember that (momentum) = (mass) $\times$ (velocity) = mv, so another way to write the quantum rule is

$$v = \frac{nh}{2mL}$$

and the kinetic energy of the electron or other small object moving back and forth along a line of length L (or bouncing between the ends of a tube) is

$$KE = \frac{1}{2}mv^2 = \frac{h^2}{8m}\frac{n^2}{L^2}.$$

Thus we have the remarkable result that only certain discrete energies of motion are allowed by the quantum rule. The possible energy is determined by n^2, with n an integer, and the electron just cannot move with any in-between energy if it is confined to that tube of a given length L. There is a sort of "ladder" or "spectrum" of allowed energies for $n = 1, 2, 3$, etc., with the rungs of the ladder getting farther and farther apart as we climb up it, as in Fig. A4–10. The energy difference between the first rung and the second, for example, is $3h^2/(8mL^2)$ and we see that this gets rapidly bigger as L gets smaller. This is a very general characteristic of these submicroscopic systems: the smaller the system, the larger the energy jump between the allowed "states" of motion. Another way of thinking of it is this: when you squeeze a particle into a very small space, you increase its kinetic energy (if n stays the same). Thus nuclei, which are much smaller than atoms, involve much greater energies.

The back-and-forth motion of an electron along a line may be described in terms of a potential energy depending on its position. It might have a PE similar to that of a ball rolling down into a valley and

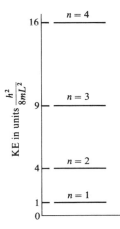

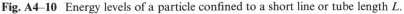

Fig. A4–10 Energy levels of a particle confined to a short line or tube length L.

up the other side, as in Fig. A4–11. If the valley were flat-bottomed with a cliff at each side, the ball would bounce between the cliffs. The electron in a tube of length L is like that; the potential suddenly rises to a very high value and we say we have a "square-well" potential or a "hard" valley. The rapid increase in the size of the jumps in the ladder of Fig. A4–10 is for a square-well potential as in Fig. A4–11(b). With a potential shaped like the curve in Fig. A4–11(a) the ladder of allowed energies is quite different. In such a "soft" valley, as n becomes larger the KE becomes larger, but this lengthens the effective length L over which the electron can move. Longer L means smaller KE, as we have seen, so the KE does not jump to as high a value as it would if L were constant. Thus the jumps between energy levels do not increase as they do in Fig. A4–11(b). In a potential that looks like a valley with sides gradually getting flatter to approach a plateau, the higher energy jumps become smaller and smaller for higher n and higher energy, as suggested by the horizontal lines in Fig. A4–11(a).

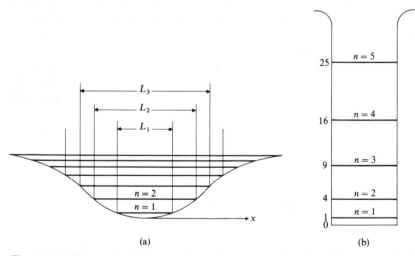

Fig. A4–11 Energy levels in a gently sloping potential well and in a steep-sided well.

Quantized Angular Momentum

Now think of a tiny particle moving around a circular path of radius r like a submicroscopic version of a ball swinging around on a string.

The distance it has to go before it completes a round trip and gets back to where it started is $2\pi r$, and the quantum rule now reads

$$(\text{momentum}) \times 2\pi r = lh.$$

It is conventional to call the integer l rather than n when we are talking about rotational motion.

As we have seen, angular momentum is an important quantity in this kind of physics. It represents a tendency for things to keep on going round and round, in the same way that ordinary momentum represents a tendency to keep on going in a straight line. For circular motion (in which v is at right angles to r), angular momentum is defined as (mass) $\times$ (velocity) $\times$ (radius) or simply (momentum) $\times$ r and from the last equation we have

$$(\text{angular momentum}) = lh/2\pi.$$

The same equation applies for the angular momentum in other types of motion around an attracting center, such as motion in an elliptical orbit. Thus angular momentum is quantized with a "quantum number" l.

The remarkable thing about this result is that it means that the electron can circulate not just with any angular momentum, but with only very special discrete values given by the integers l. There are thus different allowed states of motion, a state with $l = 1$, another with $l = 2$, etc. These possible states of motion have different energies. If we consider only those states in which the motion may be described as approximately circular, in a relatively simple "hydrogen-like" atom or ion that has only one electron but nuclear charge Ze, the possible energies are

$$E = -\text{ const } \frac{Z^2}{l^2}$$

and the corresponding radius of the orbit, for each energy, is

$$r = \text{const}' \frac{l^2}{Z}$$

(where const' is a different constant whose value need not concern us here). These expressions are derived in Appendix 7.

Quantizing Radial Motion

The elliptical motion of a comet may be described by specifying just how the direction or the angle of the radius vector changes with time and also specifying how the radius r itself changes. We may think of the motion as composed of both angular and radial motion. In the atom the angular motion is characterized by the quantization of angular momentum as we have just seen. To this we must add the fact that the radial motion is also quantized. We have seen in the case of an elliptical orbit how the radial part of the motion takes the object back and forth between a maximum radius r_{max} and a minimum radius r_{min}, that is between the outer and inner circles drawn in Fig. A4–3.

This purely radial part of the motion is something like an electron swinging back and forth across a "soft" potential valley. On the outer part of the motion, the gravitational force F_g gradually becomes more than enough to turn the motion around, $F_g > F_c$, and the radial motion turns around from outward to inward. At the inner end it is the other way around, again a gradual or "soft" change. Consider all possible motions with a certain value of the angular momentum, let us say with $l = 2$, for example. The minimum amount of radial motion is given by $n = 1$. This corresponds to a nearly circular orbit—we might say a fuzzy circle, with a little radial motion. There are other possible states of motion with $n = 2$, $n = 3$, etc. (all still with $l = 2$), and these correspond to longer and longer elliptical motions (still fuzzy). Their energies are something like those of Fig. A4–11(a), more crowded together as one goes up in n.

Consider now all energy states with one value of n, for example $n = 1$, and different l. The lowest one of these has $l = 0$, a state with no angular momentum, whose fuzzy motion goes directly in to the nucleus and out again. The motion of an electron in this state is all radial (none sideways) but so fuzzy that we cannot possibly know what radius it moves along. In effect, it fills out a fuzzy sphere. This electron is on the average so close to the nucleus that it has a very low energy, very deeply negative. With $l = 1$, $l = 2$, etc. (all with $n = 1$), we have nearly circular orbits, each one larger than the one before and separated by smaller and smaller steps as we go up the ladder, about as suggested by the formula $E = -$ const Z^2/l^2 already discussed.

Thus there are two separate reasons why the possible energies of an electron in an atom bunch together in energy as we go up toward zero

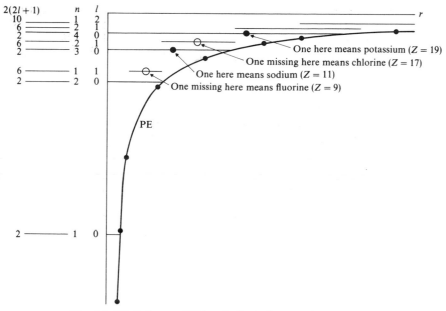

Fig. A4–12 Scheme of filling shells to form various atoms.

energy. With both reasons operative, they bunch together very closely indeed and seem to be crowded up toward zero as a ceiling.

Shell Structure of Atoms

Now we have seen that the quantum rule can be applied to show that angular motion in an orbit is quantized and that the radial motion is also quantized. These orbits lie approximately in a flat plane, so that they may be drawn on a piece of paper, so to speak. A third application of the quantum rule shows that the plane of the orbit can have only certain orientations in space, a larger number of orientations for a large value of the angular quantum number l than for a small one, namely $(2l + 1)$ orientations for each l.

There is one more important rule governing the behavior of electrons in an atom, and we may describe it as a sort of "buddy system." It states that no more than two electrons may have the same state of motion. That is, only two electrons can move in the atom with the same values of the quantum numbers n and l and a given orientation of the orbit. This means that, if we imagine a game of atom building, we can put in an atom as many as $2(2l + 1)$ electrons with the same values of n and l, but no more.

The uncertainty of position that goes along with the quantum rule means that the orbit of an electron is not quite confined to a flat plane. Instead, the plane is thickened into a sort of slab with fuzzy edges, thickened only a little for large angular momentum l and quite a lot for small l. When an atom contains all the $2(2l + 1)$ electrons to fill the "shell" corresponding to a given n and l and all the orientations are occupied, the result is a spherically symmetric distribution and the atom appears to be round like a fuzzy ball.

In Fig. A4–6 we saw how to represent an elliptical orbit by a horizontal line over the potential curve and extending from r_{min} to r_{max} at a level to represent the total energy. The outer end at r_{max} does not quite reach the PE curve and this indicates that at the outer end of the elliptical motion there is still some KE in addition to the PE. The states with no angular momentum and $l = 0$ have no such KE at the outer end of the motion, and the line for them, if drawn in a graph like Fig. A4–6, would extend out to the PE curve.

In these terms, then, and as the result of all this, the scheme by which atoms are built up is shown in Fig. A4–12. The simplest atom, hydrogen, for example, has a nucleus with $Z = 1$ in the lowest energy state with $n = 1$ and $l = 0$. If it were given some extra energy, for example by bouncing electrons off it as happens in a discharge tube such as a neon sign, it could exist in one of the higher energy states for a short time before getting rid of the extra energy by emitting light and settling down into the lowest state again. More is said about excited states of hydrogen in Appendix 7.

With $Z = 2$ there are two electrons to make a helium atom and they can both be put in the lowest state. Here we have the first example of a "closed shell": an atom with as many electrons as are allowed in the state that thus is filled. This makes it a definitely spherical atom with a large energy jump to the next allowed state so it is not easily disturbed and is chemically inert. With a slightly heavier nucleus having $Z = 3$, the third electron is put into the state with $n = 2$ and $l = 0$, the next higher state in energy.

Let us now go on to the atom with $Z = 10$. In addition to the four electrons that fill the two lowest energy states, we now have six electrons to fill the state with $n = 1$, $l = 1$ according to the rule that each such group of states can hold $2(2l + 1)$ electrons. Here again we have a closed shell and an inert atom. This atom is neon which like helium does not combine with other atoms to form molecules. With $Z = 11$ we have one additional electron in the state $n = 3$, $l = 0$. This is the chemically very active atom sodium.

And so it goes. If with $Z = 19$ we add another eight electrons, we have filled the next two shells and have one electron left over with $n = 4$, $l = 0$. This makes potassium which is chemically extremely similar to sodium. Its outer electron still has $l = 0$ but differs merely by having the next higher value of n. Similarly, fluorine with $Z = 9$ and chlorine with $Z = 17$, each lacking one electron in an $l = 1$ shell, are chemically extremely similar to each other. Thus the similarities of the orbits in these successive shells lead to periodically recurring characteristics of the atoms as we come to heavier and heavier atoms. This is usually described in terms of the "periodic table of the elements."

Actually, Fig. A4–12 must be taken "with a grain of salt." This shape of the potential curve naturally varies from atom to atom because it is caused by the attraction of different nuclear charges Ze modified by the repulsion of different numbers of electrons. Thus the last energy level filled in sodium should be about as far below zero as is the last energy shell in potassium, if we were to redraw Fig. A4–12 for each kind of atom. Appendix 9 tells how the various shells of electrons modify the shape of the potential curve and how it is much steeper at the center than in the case of hydrogen so that the inner orbits are very small indeed.

Thus our picture of an atom is of a central nucleus surrounded by concentric closed shells of electrons, slightly interpenetrating one another but of successively larger average radii, and in most cases one or several electrons not in closed shells and largely outside the outermost closed shell.

We have said that the total energy of an electron in an atom is negative. This means that if we could harness it we could make the electron do work on something outside the atom while we let the electron fall into place in the atom from a great distance. Conversely, it means that we have to do work on the atom to take an electron out of it, perhaps by throwing a projectile (such as another electron) at it. We speak of the electron in the atom as being "bound" by as much energy as it would take to remove it.

Energies of electrons in atoms (or nucleons in nuclei) are measured in terms of a unit of energy known as the electron volt, eV. We have tried to avoid overuse of units, but we need this one. A flashlight-battery cell, because of charges on its terminals, would exert a force on an electron held between its terminals. Correspondingly the electron (or any charge) near one terminal has a different potential energy from that near the other. The potential difference between the terminals is $1\frac{1}{2}$ volts, and the work we would have to do on the electron to move it from the positive to the negative terminal is $1\frac{1}{2}$ electron volts. That's not very much work, but the electron volt is a convenient unit of energy in the small world of the atom. (One eV = 1.6×10^{-19} kilogram meter^2/sec^2.)

The electron of a hydrogen atom in its lowest state ($l = 1$, $n = 1$) is bound by 13.6 electron volts, or 13.6 eV. That is, for this state $E = -13.6$ eV. The binding energies of a few other atoms having just one electron outside closed shells are shown in the Table A4–2.

TABLE A4–2
Binding energies of some simple atoms

| Atom | Z | Outer electron | | Binding energy |
		l	n	
Lithium	3	0	1	6.9 eV
Sodium	11	0	1	5.1 eV
Aluminum	13	1	2	6.0 eV
Cesium	55	0	5	3.9 eV

Consider now an atom with only closed shells of electrons, such as neon, with $Z = 10$. The outer closed shell contains 6 electrons, each with $l = 1$ and rather little radial motion, corresponding to $n = 1$. The two smaller shells inside have 4 electrons (all with $l = 0$), and these are quite effective in shielding the outer-shell electrons from the nucleus (cancelling out part of its attraction with their repulsion), so they bring the "effective nuclear charge" down from 10 to about 6. Thus an electron in the outer shell feels the attraction by a charge of about 6e from that inside part of the atom (which includes the nucleus), and feels a repulsion from the other 5 outer-shell electrons. However, they do not push as effectively as if they were near the center. Those alongside are pushing sideways rather than outward, and those on the opposite side of the atom are pushing outward but from far away. Thus the outer-shell electrons are only about 40 percent efficient in shielding one of their own number from the nucleus, or 5 of them count for about a charge of $-2e$ at the center. Each of the outer-shell electrons in neon is, then, held in about as strongly as if it were held by a nucleus with $Z = 4$. Its binding energy is 21.5 eV, much greater than that for the outer electron in the next element, sodium, with $Z = 11$. The element before neon, fluorine with $Z = 9$, has an "almost-closed" outer shell of 5 electrons, and for each of them the binding energy is 17.3 eV, again quite large and for the same reason. Another electron, if available, could stick in the vacancy in that shell, to make the neutral atom into a "negative fluorine ion," but then each electron would be repelled by one more electron and would be less tightly bound.

Thus we see how it is that an electron in the outer closed shell or almost-closed shell of an atom is bound by something like three times as strong a binding energy as is a single electron outside closed shells. Closed shells are so much favored energetically that a neutral atom with an almost-closed shell may easily become a negative ion by having an extra electron fill the vacancy.

APPENDIX 5   Period of Circular Orbit

The force characteristic of uniform circular motion,

$$F_c = mv^2/r,$$

and the inverse-square force on a body in orbit about the earth due to the earth's gravitational pull,

$$F_g = GMm/r^2,$$

are equal to each other, $F_g = F_c$, when

$$v = \sqrt{GM}/\sqrt{r}.$$

The circumference of a circular orbit is $2\pi r$ and the time taken to go around it is the distance divided by the velocity or

$$t = 2\pi r/v = \left(2\pi/\sqrt{GM}\right)r\sqrt{r} - \text{const } r\sqrt{r}.$$

For planets in approximately circular orbits around the sun, the same proportionality applies although the constant is a different one, and results in the fact that, for example, the period of Neptune, with a radius of 30 times the radius of the earth's orbit has a period of $30\sqrt{30}$ years $= 165$ years as compared with one year for the earth (Fig. A5–1).

Sun

Earth
(1 year)

Neptune
(165 years)

Fig. A5–1   Orbital of the earth and the outer major planet Neptune.

APPENDIX 6 Stability of Orbits

We consider the motion of a comet in the long elliptical orbit shown in Fig. A4–3 on page 265. The sun is at the point near the left hand of the ellipse from which the position vectors are drawn, and attracts the comet inward along the radius toward that point with an inverse-square force. There are two points at which the velocity is at right angles to the radius, the point where the radius r is longest at the right-hand end in the figure, and the point where radius r'' is smallest at the left end. We wish to show that the velocity actually does change so much between these two points that it is reasonable that the comet should fall inside the large tangent circle at the right end and then fly away outside of the small tangent circle at the left end.

At either one of these points we may think of the same two inward forces considered for a circular orbit on the previous page and may compare them with each other. One is the actual gravitational force

$$F_g = GMm/r^2.$$

The other is here a hypothetical force that would be required to make the particle continue on the dotted circular path in Fig. A4–3, starting

Fig. A6–1 Velocity is composed of one vector perpendicular to the radius and one parallel to the radius.

with its actual position and velocity.* This force characteristic of circular motion is

$$F_c = mv^2/r.$$

At the right-hand end, the elliptical motion turns inside the circle, meaning that gravitation is too strong to permit the circular motion, if

$$F_g > F_c.$$

In order to keep track of how much the velocity increases from here on, we can make use of the idea of angular momentum which is an important quantity in this kind of physics and in atoms and nuclei. As has been said, angular momentum is similarly a measure of the tendency of a body to keep on rotating about some fixed point. A body moving in a circular orbit with radius r and velocity v has an angular momentum (about the center of the circle)

$$L = rmv$$

or (radius) × (momentum). But when the body is moving with its velocity not at right angles to the radius vector, as at most of the points in an elliptical orbit, the velocity vector may be considered to be the sum of two parts or components, a vector along the direction of the radius and a vector at right angles to the radius, as in Fig. A6–1. The angular momentum L is then

$$L = \text{(radius)} \times \text{(mass)} \times \text{(component of velocity at right angles to radius)}.$$

---

* This is a more direct way to think of the force characteristic of circular motion than in terms of the idea of centrifugal force. To think of centrifugal force (which is equal to F_c but outward) properly, one must imagine himself watching the rotating object while sitting on a merry-go-round rotating as fast as the object is going around the circle. The object then seems to stay still at one point on the merry-go-round because the actual inward force seems to be balanced by an imaginary outward centrifugal force. As another way to think about the motion at the inner and outer points on the ellipse, one may consider whether the gravitational force is greater or less than the centrifugal force. Centrifugal force is a useful idea for understanding cream separators or, in the preparation of nuclear materials, centrifugal isotope separators in which the material is whirling in a rotating vessel and feels the "centrifugal force" pushing it toward the outer edge. The parts of the material with the greatest mass then tend to collect on the outside.

The most interesting property of angular momentum is that, when the only force on a body is directed toward (or away from) a fixed point, the angular momentum of the body about that point is a constant.

Now, when we compare F_c and F_g at the outermost and innermost points of the ellipse, these are just the points where v is at right angles to r so, for these special points only, $L = rmv$ or $v = L/mr$. Thus at these two points we can put this expression for v in $F_c = mv^2/r$ and can compare

$$F_c = (L^2/m)/r^3$$

with

$$F_g = \text{const}/r^2.$$

Since the angular momentum L is conserved, these two expressions vary only with r. Since they contain different powers of r, if we plot them as curves on a graph one curve is steeper than the other and they cross at only one point, as in Fig. A6–2. For example, if we reduce the radius from r to $\frac{1}{2}r$, we find that F_g containing $1/r^2$ is increased by a factor 4 while F_c, containing $1/r^3$, is increased by a factor 8, so F_c is steeper than F_g. The radius, which we may call r_c, where the curves cross and where $F_g = F_c$, is the radius at which a circular orbit is

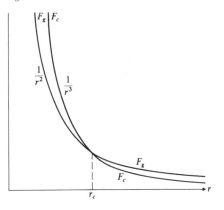

Fig. A6–2 Orbits are stable because the curves for the actual (or gravitational) force F_g and the force F_c required for a circular orbit (with a certain angular momentum) cross each other.

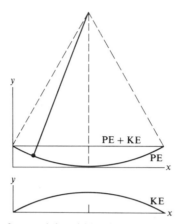

Fig. A6–3 Variation of potential and kinetic energies of a swinging pendulum.

possible with this angular momentum. An elliptical orbit swings between a maximum radius $r_{max} > r_c$ where the less steep curve is higher, $F_g > F_c$ to start the inward motion, and a minimum radius $r_{min} < r_c$ where the curves have crossed over and $F_g < F_c$. Thus we see that F_c, because it contains both v^2 and $1/r$ and because of the conservation of angular momentum, does indeed vary more rapidly than does F_g, so that it can be greater at one end of the radial motion and less at the other, and the motion can continually swing back and forth from one end to the other. A more direct way to say the same thing is that the ratio F_g/F_c is proportional to r. In the elliptical motion the ratio becomes both greater and less than one as r varies. This is the basic reason for the intrinsic stability of planetary orbits.

Even if the attracting force were somewhat different from an inverse-square force, this basic stability of the radial motion, varying between inner and outer limits, would persist. But the repetition of the motion over and over again in an exactly elliptical orbit occurs only for an exactly inverse-square force. With a slightly different law of force, if we again trace the motion inward starting from the outermost point, the next outward path would not join at the starting point. Instead, the motion would trace a series of approximate ellipses at different positions, something like the petals of a daisy. So we see that nature, in providing inverse-square forces, seems to have tailored the forces carefully to make closed orbits!

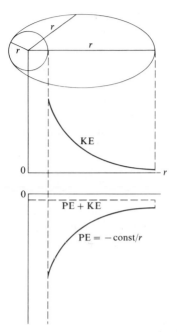

Fig. A6–4 Variation of potential and kinetic energies in an elliptical orbit.

As a comet in an elliptical orbit falls inward closer to the sun, its *KE* increases as its *PE* decreases. In this respect it can again be compared with the case of a simple pendulum. Here the path of the pendulum bob can serve as a graph of the potential energy, as in Fig. A6–3. The *KE* is zero when the *PE* is maximum at the end of the swing, and the *KE* is a maximum at the center, as shown in the same figure. The *KE* follows a curve given by an inverted portion of a circle so that the sum $E = KE + PE$ is constant.

The similar situation for a comet following an elliptical path is shown in Fig. A6–4. Corresponding to the inverse-square law of force, the potential energy varies as the inverse first power of the radius, that is,

$$PE = - \text{const}/r.$$

As *r* gets very large, the *PE* varies very little and the curve is almost flat. The *PE* is, as we have seen, conveniently defined so that it becomes zero as r becomes extremely large. The body is attracted toward positions of lower potential energy, so the PE at smaller radii is negative, as indicated in Fig. A6–4. The curve for the KE in an elliptical orbit, as a function of the distance r from the sun, curves the other way so that the sum $KE + PE$ is constant.

Thus we can think of the stability of the elliptical path as associated with the fact that part of the energy swings back and forth between KE and PE, as in a pendulum, though in the orbit neither goes to zero. For motion in a long ellipse, the KE and PE swing between wide limits while their sum remains constant, but in a nearly circular orbit, like the earth's, they swing between narrow limits.

The total energy of a body in an elliptical orbit is negative, as indicated in Fig. A6–4. One would have to add energy to it to get it up to zero energy—enough energy for the particle to go out to a very large radius, to infinity. The fact that the total energy of the solar system or of an atom or nucleus is negative then means that energy would be required to disassemble it and leave the constituent parts very far apart and at rest, that is, with zero total energy.

The inverse-square force is special also in leading to a simple relationship between KE and PE. This is easiest to show for a circular orbit under the influence of an inverse-square force. In this case the potential energy is

$$PE = -\mathrm{const}/r$$

corresponding to the force exerted, for example, by gravity:

$$F = \mathrm{const}/r^2 = -PE/r$$
$$= F_c = mv^2/r.$$

The second line comes from requiring that F be just great enough to hold m in a circular orbit. Since $KE = \frac{1}{2}mv^2$, from the last term of each line we have

$$PE = -mv^2 = -2 \times KE$$

and thus

$$E = PE + KE = -2 \times KE + KE = -KE = \tfrac{1}{2}PE.$$

The total energy is then negative and equal in magnitude to the kinetic energy, the potential energy being negative and twice as great.

For an elliptical orbit the situation is much more complicated, with *KE* and *PE* rapidly varying as we have seen, but it turns out that the same relation holds if we speak of average *KE* and average *PE*, averaged over time. In a long ellipse the body spends most of the time in the outer part where it moves slowly and the average is close to values found out there.

APPENDIX 7 Allowed Energies in a One-Electron Atom

As we have seen in Appendix 4, only certain discrete states of motion are permitted by the quantization of angular momentum:

$$\text{ang. mom.} = lh/2\pi, \qquad l = 0, 1, 2, \ldots$$

where h is Planck's constant and l is the angular momentum quantum number. While there are states approximating elliptical orbits as well as circular orbits, we will have a pretty good idea of how the energies are distributed if we consider only the simple states whose orbits are approximately circular, and consider them as exactly circular orbits even though they are actually "fuzzy." The force of attraction Ze^2/r^2 between the nucleus and the electron must then be just equal to the force characteristic of circular motion $F_c = mv^2/r$.

Then we have

$$\frac{Ze^2}{r^2} = m\frac{v^2}{r}. \qquad \text{Thus} \qquad r = \frac{Ze^2}{mv^2}.$$

When put in the expression for angular momentum, this gives us the relations

$$rmv = \frac{Ze^2}{v} = l\frac{h}{2\pi} \qquad \text{or} \qquad v = \frac{2\pi}{h}\frac{Ze^2}{l}.$$

The energy of the electron in this state of motion, attracted toward the nucleus by an inverse-square force, is

$$E = -KE = -\frac{1}{2}mv^2 = -\text{const}\frac{Z^2}{l^2}$$

[where, incidentally, $\text{const} = \dfrac{m}{2}\left(\dfrac{2\pi}{h}\right)^2 e^4$]. The expression for r, $r = Ze^2/mv^2$, has mv^2 down in the denominator, which means that r is small when the magnitude of E (or of KE) is large. This is because the electron is deep in the sharp depression of the PE graph ($PE = -Ze^2/r$), pulled down in close to the nucleus when r is small. Explicitly,

$$r = \text{const}'\,\frac{l^2}{Z}.$$

[This "constant prime" is simply a different constant that happens to be equal to $(h/2\pi e)^2/m$ but this need not concern us.]

The total energy is negative, being the sum of a negative potential energy and a somewhat smaller positive kinetic energy. We see that this negative total energy becomes large in magnitude for large Z (when a strong force pulls the atom together into a small space) and small for large angular momentum quantum number l (which is characteristic of large orbits). For the radius the formula goes the other way: the radius is large for large l and small for large nuclear charge Z.

A hydrogen atom has a sequence of states of this type (roughly corresponding to circular orbits) for the various values of $l = 1, 2, 3$, etc., and their energies are arranged in a spectrum that looks not like an ordinary ladder but like a ladder with the rungs crowded together near the top, as in the left-hand part of Fig. A7–1. It is sometimes

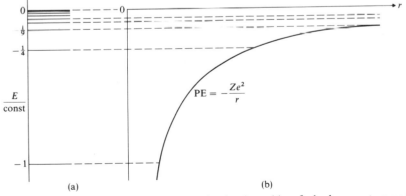

Fig. A7–1 Energy spectrum of the nearly circular orbits of a hydrogen atom and an indication of how they fit into the potential well.

convenient to plot these energies as short lines inside the "potential well" formed by the graph of the *PE*. The right-hand side of Fig. A7–1 shows this. This graph shows that the radius is small for a state with energy deep down at large negative energy, held in by the potential well (or the corresponding force).

APPENDIX 8 The Wave Picture of Atoms

A wave is a periodic variation of something. Water waves are perhaps the most familiar kind. Figure A8–1 might represent a snapshot of a cross section of the surface of the water in a tank. The horizontal line represents the average height of the surface. At some points the wave is above the average or positive, and at others below the zero line or negative. In the case of a simple "standing wave" that vibrates up and down but does not move to right or left, the level remains at zero at both ends of the tank. Let us speak of each crest or trough of the water as a wavelet. Thus in Fig. A8–1 there are five wavelets, but if we agitate the water differently there might be six or seven. The wavelength is the distance after which the wiggle starts to repeat itself, as indicated in the figure, so a wavelength consists of two wavelets.

Radio waves are similar periodic variations of certain electric properties in space, and light rays, x-rays, and gamma rays are essentially the same thing with progressively very much shorter wavelengths, extending far down into submicroscopic dimensions. This fact can be demonstrated in the case of light rays and x-rays by the peculiar patterns they make when they are reflected either by mirrors with evenly spaced scratches on them, or, in the case of x-rays, by crystals in which rows of atoms take the place of the regularly spaced scratches.

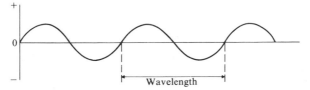

Fig. A8–1 The wave associated with an electron moving along one direction.

A remarkable fact that is at the basis of our understanding of atoms is that electrons, when bounced off of crystals, make characteristic patterns very similar to those made by x-rays, which are waves. This indicates that the tiny electrically charged particle that we call an electron must be guided by a wave accompanying it that tells it where it can go. This is not the wave of anything physical, but of a mathematical function that is periodically positive and negative and has to behave properly at the ends of the motion if the motion is confined. For example, the wave guiding the behavior of a tiny particle bouncing back and forth between the ends of a very small tube has to go down to zero at the ends of the tube just as does the "standing" water wave at the ends of the tank.

Now this mysterious wave, which has to be accepted as a fact of life in the submicroscopic world, is associated with the particle and its motion through the relation

$$\text{momentum} = \frac{h}{\text{wavelength}}$$

or the other way around,

$$\text{wavelength} = \frac{h}{\text{momentum}}.$$

where h is Planck's constant.

In Fig. A8–1 one sees that the length L of the tube must contain an integral number of half wavelengths, that is

$$n \times \frac{\text{wavelength}}{2} = L \quad \text{or} \quad \text{wavelength} = \frac{h}{\text{momentum}} = \frac{2L}{n}$$

which is the same as the quantum rule used in the text,

$$\text{momentum} \times 2L = nh$$

The description of the allowed motions in terms of waves cannot be given without the use of rather demanding mathematics, such as differential equations. Here we can only give a descriptive picture of the atom in terms of these waves. The wave picture modifies classical mechanics in a way that makes no difference for the motions of large bodies. For example, the number of wavelengths describing the moon's orbit is a number so astronomically large that the fact that it is an integer makes no difference at all. But it makes a great deal of difference in small systems like atoms and nuclei. In particular it permits only certain allowed energies of states of motion.

When a particle according to the macroscopic rules is moving in an orbit such that it comes back to where it started from and starts over, the wave guiding its motion must be a single wave and must join up with itself smoothly, not leaving any loose ends. The bouncing in the tube is one case of this; there must be an integral number of wavelengths for the round trip. For the case of an electron in a circular orbit, the smooth joining of the wave is indicated in Fig. A8–2. Again this is equivalent to the quantum rule in the text, this time the quantization of angular momentum, since it says

$$\text{circumference} = 2\pi r = n \text{ wavelengths} = \frac{nh}{\text{momentum}}$$

or

$$\text{angular momentum} = r \times \text{momentum} = nh/2\pi.$$

We may think of such waves going around circles of various radii r but all with the same number of waves n and the same angular momentum. The smaller the radius the shorter the wavelength and the larger the momentum. If these waves at various values of r are all in step with one another, their crests will be lined up to look like the spokes

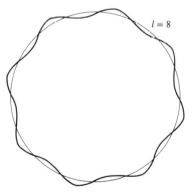

Fig. A8–2 The wave associated with an electron in an approximately circular orbit.

of a wheel. They are thus waves in angle, describing angular momentum.

In the classical motion of an elliptical orbit, the moving object has not only angular momentum but momentum inward and outward along the radius—radial momentum. The radial momentum is governed by the wave picture, too. It is possible for a wave on a surface to undulate in two directions at once; think of two sets of ripples on a pond crossing one another and making a small checkerboard pattern where they cross. As the elliptical motion moves in and out along the radius, the momentum changes drastically with the change of potential energy and the local wavelength of the associated wave changes accordingly. Thus the radial wave is a distorted wave, not a highly regular one that repeats itself over and over like a wave on a lake.

There is no way for the two ends of the radial wave to fit together the way the angular wave does. Instead, it fades off to zero at large distances. A typical radial wave is shown in Fig. A8–3. After going

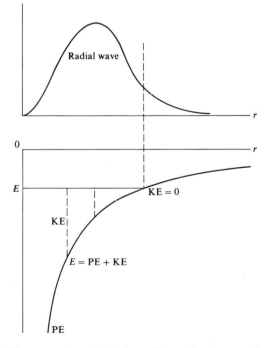

Fig. A8–3 Wave associated with the radial motion in a nearly circular orbit.

over a hump, it tapers off to zero at large r. In the lower part of the figure the horizontal line E indicating the total energy is shown extending out until it meets the PE curve. At any radius the distance from this level down to the PE curve represents the KE, since $E = PE + KE$. Where the line E meets the PE curve the KE is zero, and here the radius is as large as is classically possible for a particle of this energy.

In classical mechanics, if we imagine that we send a small ball with total energy E rolling up a hill shaped like the PE curve, it would stop when it reaches the level E, and for an instant it would have $KE = 0$ before it starts rolling back down the hill. In wave mechanics, the wave penetrates out beyond this point but gradually tapers off toward zero. This means that the motion of the particle can occasionally extend out into that region where classical mechanics would not let it go. This is associated with the idea that we cannot be sure just where a particle is in the small world of the atom.

One important use of the wave is to determine quantized angular momenta and energies, because the wave must join itself when the motion goes around a closed circuit. Another important use of the wave is to indicate what part of its time the electron spends at various places. To state the rule for this, we should first plot a curve that is the square of the height of the wave, a curve that is relatively smaller where the wave is already small, and that peaks up more sharply at the hump, as in Fig. A8–4. The rule is then that the fraction of its time that the electron spends in the immediate vicinity of any radius is proportional to the height of the "wave-squared" curve at that r.

For example, if the maximum of the wave is at r_m and it comes down to half its maximum value at r_h, then the electron spends only

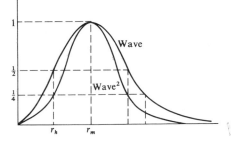

Fig. A8–4 The square of the height of the wave, which tells where the electron is apt to be, is more sharply peaked than the radial wave itself.

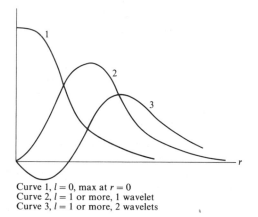

Curve 1, $l = 0$, max at $r = 0$
Curve 2, $l = 1$ or more, 1 wavelet
Curve 3, $l = 1$ or more, 2 wavelets

Fig. A8–5 Radial waves for various orbits.

$\frac{1}{4}$ as much time near r_h as near r_m. In this revised view, since it is not possible to know exactly where the electron is, we should not have drawn the circular orbit as a narrow line, but rather as a broad fuzzy band. The wave^2 curve in Fig. A8–4 shows how fuzzy the band indicating the orbit should be. The idea of a circular orbit is helpful in understanding the structure and the energies of this "circular-orbit" group of states of motion, but the circular-orbit picture it brings to mind is only approximately true.

In a similar way there are other possible states of motion of the electron that correspond approximately to elongated elliptical orbits. In an elliptical orbit an electron has not only the perpendicular-to-radius or angular motion that we have discussed but also has momentum along the radius. Even the motion in an approximately circular orbit is not quite circular and has some momentum along the radius, but only a little and its wavelength in this direction is long, as in curve 2 of Fig. A8–5. A really elongated ellipse has more radial momentum and a shorter wavelength in the radial direction, as in curve 3 of Fig. A8–5. This curve crosses the axis once, passing through zero from negative to positive values, and has two wavelets. There is a whole group of states with two radial wavelets, each having a different value of angular momentum, given by $l = 1, 2, 3$, etc. Corresponding to still more elongated ellipses, there are other groups of states with even more wavelets. For a given value of angular momentum more radial wavelets mean higher energy (higher on the energy-level ladder, though still negative).

If the electron has some angular momentum it cannot be at the center, very close to the tiny nucleus, because angular momentum is radius times a component of momentum. (When $r = 0$, angular momentum $= 0$.) Curves 2 and 3 represent states with $l = 1$ or more, and for this reason they start up from zero at $r = 0$. There also exist states of motion of the electron with $l = 0$, that is, with no angular momentum but only radial momentum. Curve 1 is the radial wave for such a state. It has only one wavelet, at the center, but there are other $l = 0$ states that have additional radial wavelets and higher energy.

Thus altogether there are more allowed states than indicated on the "ladder" of Fig. A7–1, but the energy jump from the lowest state to the next lowest is still the largest jump, and the jumps between energies become gradually smaller after that.

The purpose of all these curves and this discussion of angular momentum and energy states and waves is to convey a fairly accurate idea of what an atom is like. A description of the atom (and of molecules and nuclei) without these ideas is apt to be misleading. Now, with this more elaborate concept of what is meant when we describe an orbit, we can go on and construct a picture of more complicated atoms in Appendix 9. The quantum rule used there is simpler and gives a fairly good approximation to the results of the wave picture and some idea of what an atom is like.

Closed Shells and Potential
Energy in an Atom

When an atomic shell is "occupied" by its full quota of $2(2l + 1)$ electrons, these electrons form a completely round or spherical assembly. All these electrons have the same radial wave and thus the same average radius and the same fuzziness about this average r. The orbits of the various electrons have different orientations in space, in such a way that a picture or sculptural model showing where the electrons are apt to be would resemble a round ball of thick sponge rubber, hollow in the middle. The distribution of the electrons is then said to have spherical symmetry. Their angular momenta are also arranged so that they are no different seen from one direction than from any other. If we could look in along the x-axis, for example, we would see just as many electrons circulating clockwise as counterclockwise. In this sense the angular motions just cancel one another out and we say that this group of electrons has no total angular momentum. Because of this complete spherical symmetry of the electrons in the state, we speak of them as forming a closed shell.

A charged sphere exerts on a point charge outside of it a force directed exactly along the radius between centers. In fact, the inverse-square law of electric force works when Q_1 and Q_2 are both uniformly charged spheres, or between a sphere and a point charge, or between two point charges. (This is not true of irregularly shaped bodies fairly close together.)

If an atom has one or more closed shells, plus one other electron, the total force on that electron is along the radius. This means that the conservation of angular momentum still applies; there is no force like a force on a crank to change the rotation. Then most of our discussion of angular momentum is still relevant; in particular, there are still states of motion of the single electron with $l = 0, 1, 2$, etc., and for each of them there are radial motions with $n = 1$, $n = 2$, etc. However, the force is no longer an inverse-square force, and the potential energy curve is more complicated than the $-e^2/r$ for hydrogen. This changes the exact shape but not the general nature of the radial waves. When the Zth electron is outside the closed shells, there are $Z - 1$ electrons

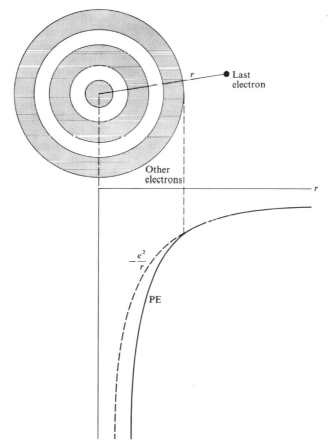

Fig. A9–1 In an atom with $Z - 1$ electrons in closed shells, the potential energy curve outside the closed shells is the same as for the hydrogen atom, but inside it is much deeper.

repelling it with a force that almost cancels out the attraction of the Z protons in the nucleus. The total force is just what it would be with just one proton at the center attracting it.

If there are several closed shells, the two electrons in the inner shell ($l = 0, n = 1$) feel the strong pull of the charge Ze on the nucleus, and they have a very small radius corresponding to the large Z. The electrons in the next shell ($l = 0, n = 2$) are partly shielded from the nucleus by the inner closed shell and have a larger average radius both for this reason and because of the greater radial momentum that tends to carry them further out. Because of the repulsion of the inner electrons the successive shells thus feel a successively weaker force pulling them toward the center and have successively larger average radii. The shells partly penetrate one another and form an atom that is spherical except possibly for the orbits of the last few electrons outside closed shells.

Thus the closed shells "shield" an outer electron from the nucleus, so to speak. If an outer electron is at large r, we still have $PE = -e^2/r$. If the electron penetrates inside the closed shells, it feels the stronger force and deeper potential of the unshielded nucleus (Fig. A9–1). It acquires so much kinetic energy that it is quickly thrown out again (somewhat like a comet zipping around the near end of a long ellipse). The low-energy orbits are already occupied by the closed-shell electrons, and the odd electron has to have a higher-energy radial wave. If it has $l = 0$, for example, it has some little short-wavelength wavelets for small r inside the closed shells and one higher hump outside, indicating that it spends almost all its time outside the closed shells where the PE is the same as for hydrogen. Thus it does not have as low an energy as the lowest state of hydrogen, in which the electron spends most of its time in the one wavelet near the nucleus. The lone electron outside closed shells again has a whole ladder of energy states, but they are not as widely separated as for hydrogen.

APPENDIX 10 Behavior of Nucleons in Nuclei

We have seen that a nucleon in a nucleus can swim through a sea of other nucleons with an average potential energy that is almost constant, bouncing back in when it gets to the edge of the nucleus because it does not have enough energy to get out. This motion suggests an "orbit" quite different from the elliptical orbits in atoms, but the motion is still governed by the same quantum rule. Because of the flat-bottomed potential the energy levels are arranged quite differently from those in atoms. Many nuclei are spherical and for them Fig. 3–12 represents a round or spherically symmetric potential in which a nucleon can coast around and around. One possible shape of orbit in such a round nucleus is a circular orbit, just as in an atom. It is possible to start an ice cube sliding so as to make it follow a circular orbit inside a fruit bowl shaped like the right side of Fig. A10–1, just as it is in the flatter dish of Fig. A4–4. But the noncircular orbits are quite different in the two cases.

Whether the orbit is circular or not, in these round nuclei as in atoms we have conservation of angular momentum with the quantized values

$$\text{ang. mom.} = lh/2\pi.$$

We again have shells. But now we have two kinds of nucleons, namely, protons and neutrons, and the rule that $2(2l + 1)$ of them can be in a shell applies to protons alone and again to neutrons alone. Thus a filled shell in the nucleus, if it is filled with both protons and neutrons, can have twice as many particles in it as can the electron shells in an atom.

Nuclei have various numbers Z of protons and N of neutrons. This shell model of nuclei works surprisingly well for all nuclei that are any-

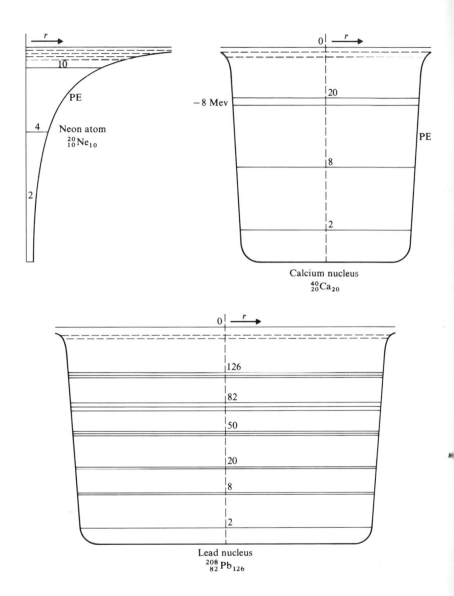

Fig. A10–1 Comparison of the energy-level patterns in a fairly light atom, a fairly light nucleus, and a heavy nucleus. The atom is of course very much larger than the nuclei and its energy scale very much smaller.

where near to having filled shells. In atoms we have seen that closed shells have a special stability—that the binding energy for an outer electron in the atom with $Z = 10$ is much greater than for $Z = 11$, for example. The same thing happens in nuclei, largely because going to the next shell means increasing kinetic energy. In nuclei the numbers Z and N corresponding to closed shells are called "magic numbers." They are 2, 8, 20, 50, 82, and 126. These are different from the corresponding numbers in atoms because the potential-energy curve is different. The ways the first few shells are filled in atoms and in nuclei are compared in Table A10–1. The main difference here between atom and nucleus is that the second $l = 0$ electron state in an atom is pulled way down in energy into the deep potential hole near the nucleus, so it is the second shell in an atom like neon, whereas this state in a nucleus has practically the same energy as the first $l = 2$ state. Having about the same energy, these two count as belonging to the same shell—a sort of double shell—when we are looking for nuclei with favorable total binding energies. That is the meaning of the brackets joining them in the table, and it accounts for the third "magic number" being 20. The explanation of the closed shells accounting for the larger "magic numbers" 50, 82, and 126 requires going a little further into details of nuclear structure than we can afford here, but this gives the general idea of how the levels bunch together and leave gaps.

Figure A10–1 compares the ladders of energy levels in an atom, in a

TABLE A10–1

Filling of shells in atoms and in nuclei

		Atoms					*Nuclei*		
l	n	*Number of electrons filling shell*	*Total number*	*Name of atom*	l	n	*Number of electrons filling shell*	*"Magic number"*	*Nucleus with N and Z "magic"*
0	1	2	2	Helium	0	1	2	2	$^4_2\text{He}_2$
0	2	2	4	Beryllium	1	1	6	8	$^{16}_8\text{O}_8$
1	1	6	10	Neon	$\left\{\begin{array}{c} 2 \\ 0 \end{array}\right.$	$\begin{array}{c} 1 \\ 2 \end{array}$	$\left.\begin{array}{c} 10 \\ 2 \end{array}\right\}$	20	$^{40}_{20}\text{Ca}_{20}$

fairly light nucleus and in a heavy nucleus which occupies a larger volume. In the atom the levels are crowded together toward the top. In the nucleus there are almost equal intervals between the levels or groups of levels. In this somewhat simplified portrayal of a nucleus the average potential energy is almost the same for a nucleon in any one of the various energy levels, because the potential energy depends so little on where the nucleon is within the nucleus. The differences in energy of the various states arise mainly from differences in kinetic energy. We note that the relative spacing of the first three levels, up to magic number 20, are similar in the light and heavy nuclei, but much more widely spaced in the lighter nucleus. This is because the lighter nucleus is smaller, and where the quantum rule applies, small size means greater energy spread between states. We saw this in the case of a particle confined in a thin tube. In a round nucleus there is an angular quantum number l and a radial quantum number n. They both arise from using the quantum rule that has the form

$$(\text{momentum}) \times (\text{path length}) = (\text{quantum number}) \times h.$$

If we compare the states in a nucleon in a large nucleus and one in a small nucleus, but with the same values of l and n, then the path length is smaller in the small nucleus and the momentum larger, so the KE is greater in the smaller nucleus. (For l, the path length is the circumference of a circle and for n it is along the radius, but both become smaller for the smaller nucleus.) This is the main reason why the energy levels in the small calcium nucleus in Fig. A10–1 are more widely spaced than those in the larger lead nucleus.

In the tube with hard ends, the kinetic energy varied as

$$KE = \text{const}\,\frac{n^2}{L^2}$$

(in which n is the quantum number for that case and L the length of the tube). In the atom the KE varies much more slowly that n^2, even more slowly than n for the radial motion, because the edges of the atom are so "soft." The sharp rise of the PE near the edge of a nucleus, as in Fig. A10–1, indicates that a nucleus is quite hard at the edges but not completely hard. For a nucleus the variation is between the other two: the energy varies about in direct proportion to l and to n. This is why the groups of energy levels are approximately evenly spaced in nuclei.

The two nuclear "potential wells" shown in Fig. A10–1 both have about the same depth—about 30 Mev—but one is larger in extent, corresponding to the larger sphere filled out by the lead nucleus. The nucleons occupy various energy levels extending down to about -30 Mev. But if we ask what is the total binding energy of the lead nucleus in terms of taking it apart, we should start with the least-bound nucleons at the top of the heap. It takes about 8 Mev to remove one of them (on the average). As we take a few of these out, the size of the resulting nuclei becomes gradually smaller, the whole ladder of energy levels moves up, so that by the time we are emptying the levels from the next group (just under the magic number 82), it has come up high enough that about 8 Mev is required, on the average, to remove a nucleon from it. By the time we get down to the calcium nucleus shown, the upper occupied level is still at about -8 Mev. Thus we can say that the average energy by which the nucleons are bound, in terms of the energy required to take the nucleus apart into its constituent nucleons, is about -8 Mev per nucleon.

Transfer of Kinetic Energy
in an Elastic Collision

When two objects collide, their velocities change and their kinetic energies generally change. If they are two chunks of lead, some energy is lost in deforming them, but if they are two elastic bodies, such as two ideal billiard balls, no energy is lost and the total kinetic energy after collision is the same as before. Some kinetic energy is transferred from one to the other. By combining conservation of kinetic energy with conservation of momentum one can calculate how much is transferred. This is important as a means of slowing down neutrons in nuclear reactors. The calculation is the same whether it be for a neutron hitting a nucleus or a billiard ball hitting a polo ball.

Consider a mass m (the billiard ball) with velocity v hitting a mass M (the polo ball) that is sitting at rest, and consider the special simple case in which it hits so squarely that the recoil is along the line of v. This is a one-dimensional problem; all motion is along the x-axis, say. After the collision the velocity of m we call v' and the velocity of M is V. The conservation of momentum then says

$$mv = mv' + MV.$$

Conservation of kinetic energy (in an elastic collision) says

$$\tfrac{1}{2}mv^2 = \tfrac{1}{2}mv'^2 + \tfrac{1}{2}MV^2.$$

First, let us get rid of one of the final velocities so as to be able to find the other. Each of these equations provides an easy way to write MV^2:

$$MV^2 = \frac{m^2}{M}(v - v')^2 = m(v^2 - v'^2).$$

The handy and easy-to-prove algebraic relation

$$(v^2 - v'^2) = (v + v')(v - v')$$

then gives

$$v + v' = \frac{m}{M}(v - v') \quad \text{or} \quad (M + m)v' = -(M - m)v.$$

Note that the mass m bounces backward (v' is negative) if M is greater than m. From this we find how the kinetic energy of the mass m after collision compares with that before collision:

$$\tfrac{1}{2}mv'^2 = \left|\frac{M - m}{M + m}\right|^2 \tfrac{1}{2}mv^2.$$

As illustrative extreme cases, note that when $m = M$, the mass m stops dead and has no KE, whereas when M is very much greater than m, (written $M \gg m$), there is almost no recoil velocity of M and m bounces backward with practically undiminished KE.

For less direct collisions from which m bounces off at an angle, the situation is somewhat similar. With $M \gg m$ there is practically no transfer of KE, but with $M = m$ the "projectile" loses quite a lot of KE but not all. As one sees from the simpler derivation, the general reason for this is that momentum is linear whereas KE is quadratic in v. With momentum in mind, it is hard to move a massive object enough to have much effect on its KE.

Time Available for Energy
Production in a Nuclear Explosion

When a plutonium sphere is at such a density as to be more than a
critical mass, it produces energy at a high rate but only for a short time
before it has overcome the inertia of its outer parts and blown itself
apart. Inertia can delay dispersal of the sphere for only a short time,
and the question is whether the time can in principle be long enough.
When the plutonium is near its maximum compression, the chain
reaction may be producing energy at a certain very high rate (which we
may call P, since rate of energy production is power). It will do so only
for a very short time, t, before it is blown apart, roughly speaking. The
total energy produced is equal to the rate of production (actually some
average rate) times the time. In symbols

$$E = Pt.$$

It seems clear that the more rapidly the energy is produced, the shorter
will be the time. That is, as P increases, t decreases. One might then
ask, does it do any good to increase the rate of production; does this
not simply decrease t so much that the product, E, is not increased?
This is a way of asking whether there is enough inertia to hold the thing
together long enough to make an explosion. Experience tells us there
is, but some people can understand what is going on a little better with a
few equations. (Of course people in the business use much more complicated equations in which the actual numbers might be important,
but here we use the equations to indicate only how one quantity varies
with another.)

One simple way to bring inertia into the discussion is to point out
that after the time t, a large part of the energy produced will have been

transformed into kinetic energy of the outer portion, say the outer half, of the sphere. This outer part (with mass M) will have a velocity v which is roughly equal to the radius, R, of the original sphere divided by t because the outer edge needs to move a distance of the "order of magnitude" R (actually somewhat less than R) before the expansion brings this sphere to its subcritical size and the chain reaction stops. We may write this

$$v = R/t$$

where R may be treated as a constant. This equation simply means "the higher the velocity, the shorter the time" before the expansion stops the chain reaction. The assumption that the energy produced is approximately equal to the kinetic energy of the outward motion at the end of time t is expressed by

$$E = \tfrac{1}{2}Mv^2 = \tfrac{1}{2}MR^2/t^2 - Pt$$

where M may be taken to be the mass of roughly the outer half of the material, also to be treated as a constant. The second equality in this string of equations is obtained by substituting for v and the third by substituting for E. By multiplying both sides by t^2 and dividing by P, this last equality can be transformed into

$$t^3 = \tfrac{1}{2}MR^2/P = \text{const}/P.$$

This equation means "the higher the rate of energy production, the shorter the time, but not very much so." For example, one would have to increase the power P by a factor 8 in order to make the time t half as long, since $(\tfrac{1}{2})^3 = \tfrac{1}{8}$. This is the crux of the matter: the time does depend upon the rate of energy production, but not very sensitively, so the chain reaction is not cut off as quickly as one might think. If we substitute this equation back into the first equation above, after taking the cube of each side of it, we obtain

$$E^3 = \text{const} \times P^2.$$

This means that E does increase with P and that very fast energy production can make a big bang in spite of the short time.

APPENDIX 13 Quantities of Energy

Energy may exist in several forms and is measured in several units, some of the units being intended for use with particular forms of energy, but all of them interchangeable as ways of expressing a given amount of energy. Some forms of energy are: potential, which may be gravitational, electric, or magnetic; kinetic; thermal; radiant; chemical. A unit introduced for gravitational energy is the foot-pound, for thermal energy the British thermal unit (Btu) and the calorie, for energy conveyed by an electric current the kilowatt hour, etc.

Energy given in one unit may be converted to other units by means of Table A13–2. Table A13–1 gives some equivalent units of power.

TABLE A13–1

Units of power

	kw	*hp*	*ft-lb/min*
Kilowatt	1	1.35×10^{-3}	44.4
Horsepower	746	1	3.3×10^{4}
Foot-pound per minute			1

TABLE A13–2

Units of energy

	kwh	Mev	amu	Btu	cal	ft-lb	erg	Ton coal	bbl oil	kg U-235	MT
Kilowatt hour	1	2.3×10^{19}		3.3×10^3			3.6×10^{13}	1.3×10^{-4}		2.8×10^{-8}	8.4×10^{-10}
Million electron volt		1					1.6×10^{-4}				
Atomic mass unit		931	1		3.6×10^{-11}						
British thermal unit	3×10^{-4}	6.8×10^{15}		1	252	778	1.05×10^{10}	3.9×10^{-8}	1.7×10^{-9}	8.6×10^{-12}	2.6×10^{-13}
Calorie				0.004	1		4.7×10^{-7}				
Foot-pound				1.3×10^{-3}		1	1.36×10^7				
Erg	2.8×10^{-14}					7.4×10^{-8}	1				
1 ton of coal	7.8×10^3			2.6×10^7				1	4.3		6.8×10^{-6}
1 barrel crude oil	1.8×10^3			5.8×10^8				0.23	1		
1 kg of U-235	3.5×10^7	8×10^{26}		1.2×10^{11}						1	0.03
1 megaton TNT	1.2×10^9									33	1

Example of use of the table: One kilowatt hour is the equivalent of the heat released by the burning of 1.3×10^{-4} tons of coal (or about 1/4 pound of coal) so one ton of coal is the equivalent of $1/1.3 \times 10^{-4}$ kwh $= 7.8 \times 10^3$ kwh.

APPENDIX 14 Numbers of Reactors

Nuclear Electric Generating Plants of the "Western" World

	Start up	Type*	Generating capacity, MW (electric)	Total production since start up, millions of MWh
United States				
Shippingport (Pa.)	1957	P	100	4.6
Dresden 1 (Ill.)	1959	B	210	11
Yankee Rowe (Mass.)	1960	P	185	12
Indian Point 1 (N.Y.)	1962	P	292	10
Big Rock Point (Mich.)	1962	B	75	2.8
Humboldt Bay (Cal.)	1963	B	65	2.8
Peach Bottom 1 (Pa.)	1966	G	40	0.7
Connecticut Yankee (Conn.)	1967	P	590	13
San Onofre 1 (Cal.)	1968	P	430	9
Oyster Creek (N.J.)	1969	B	550	5.5
Nine Mile Point (N.Y.)	1969	B	525	3
Ginna (N.Y.)	1969	P	450	3
Dresden 2 (Ill.)	1969	B	809	1.9
LaCrosse (Wisc.)	1969	B	50	0.5
Point Beach 1 (Wisc.)	1970	P	524	1.2
Millstone Point 1 (Conn.)	1969	B	682	1.0
United Kingdom				
Calder Hall			219	22
Chapelcross			228	19
Berkeley			335	19

* P, pressurized-water; B, boiling-water; G, gas-cooled, graphite-moderated.

Nuclear Electric Generating Plants of the "Western" World—*continued*

	Start up	Type*	Generating capacity, MW (electric)	Total production since start up, millions of MWh
Bradwell			374	20
Hunterston			360	18
Hinkley Point			664	21
Trawsfynydd			584	18
Dungeness			576	19
Sizewell			652	17
Oldbury			630	9
Windscale			11	1.8
Dounreay			14	0.4
Winfrith			100	1.3
France				
Marcoule			80	6
Chinon			760	14
St. Laurent-des-Faux			500	1.8
Chooz			266	2.7
Italy				
Latina			210	10
Garigliano			169	7
Trino			252	5
West Germany				
Gundremmingen			250	6
Lingen			255	4.5
MZFR			57	1
Obrigheim			345	5.5
Japan				
Tokai			166	4
Tsuruga			322	3
Mihama 1			340	1.2
Fukushima			440	0.4
Canada				
Douglas			220	2.7

Nuclear Electric Generating Plants of the "Western" World—*continued*

	Start up	Type*	Generating capacity, MW (electric)	Total production since start up, millions of MWh
Sweden				
Agesta			12	0.3
Spain				
Zorita 1			160	2.2
Garona			460	0.1
Netherlands				
Dodewaard			54	0.8
Switzerland				
Beznau			364	3.3
India				
Tarapur 1			200	
Tarapur 2			200	3.5

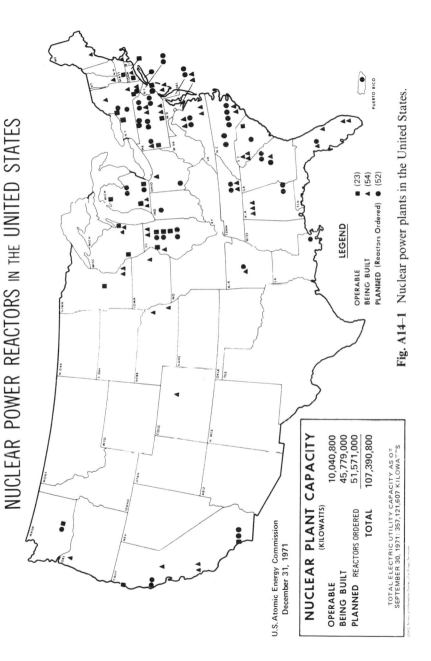

NUCLEAR POWER REACTORS IN THE UNITED STATES

U.S. Atomic Energy Commission
December 31, 1971

NUCLEAR PLANT CAPACITY
(KILOWATTS)

OPERABLE		10,040,800
BEING BUILT		45,779,000
PLANNED REACTORS ORDERED		51,571,000
TOTAL		107,390,800

TOTAL ELECTRIC UTILITY CAPACITY AS OF
SEPTEMBER 30, 1971: 357,121,607 KILOWATTS

USAEC, Technical Information Center, Oak Ridge, Tennessee

LEGEND

■ OPERABLE (23)
▲ BEING BUILT (54)
● PLANNED (Reactors Ordered) (52)

Fig. A14-1 Nuclear power plants in the United States.

Nuclear Technology, Handling,
and Safeguards of Nuclear Materials

Excerpts from important speeches and writings

These excerpts, like any excerpts, are open to the danger that "quotation out of context" may be misleading, though it is hoped that here they are not. Readers who become especially interested in some of these subjects would do well to seek out the complete texts for more complete information.

FAST BREEDER REACTORS
Glenn T. Seaborg and Justin L. Bloom
From *Scientific American* **223**, 13 (November 1970).*

The need to generate enormous additional amounts of electric power while at the same time protecting the environment is taking form as one of the major social and technological problems that our society must resolve over the next few decades.

Nuclear reactors of the breeder type hold great promise as the solution to these problems. Producing more nuclear fuel than they consume, they would make it feasible to utilize enormous quantities of low-grade uranium and thorium ores dispersed in the rocks of the earth as a source of low-cost energy for thousands of years. The United States Atomic Energy Commission, the nuclear industry and the electric utilities have mounted a large-scale effort to develop the

---

* Complete text available as a *Scientific American* offprint, San Francisco, W. H. Freeman, 1970.

technology whereby it will be possible to have a breeder reactor generating electric power on a commercial scale by 1984.

In the United States and several other countries decisions were made rather early that a fast breeder reactor cooled with liquid metal was the most attractive concept to pursue. The greater part of breeder-reactor development is now proceeding on the basis of this concept. A serious alternative effort is being pursued, chiefly by utility companies, to develop the technology of a gas-cooled fast breeder reactor using pressurized helium as the coolant. In the United States two thermal-breeder-reactor concepts operating on the thorium cycle are being developed: the light-water breeder reactor at the Bettis Atomic Power Laboratory and the molten-salt reactor at the Oak Ridge National Laboratory.

We believe breeders will result in a transition to the massive use of nuclear energy in a new economic and technological framework. The transition may be slow, and it will require the introduction of a series of innovations in the technologies of industry, agriculture and transportation; electrification of the metal and chemical industries; desalting plants; electromechanization of farms and of means of transportation; electrification of the metal and chemical industries, and more effective means for utilizing wastes. The key to these possibilities is abundant low-cost electrical energy, and the route to that is by way of the breeder reactor.

As for the environment, the extent of public concern over improving the quality of air, water and the landscape hardly needs elaboration, except for one point that is often overlooked: it will take large amounts of electrical energy to run the many kinds of purification plants that will be needed to clean up the air and water and to recycle wastes.

A related problem of equal magnitude is the rational utilization of the nation's finite reserves of coal, oil and gas. In the long term they will be far more precious as sources of organic molecules than as sources of heat. Moreover, any reduction in the consumption of organic fuels brings about a proportional reduction in air pollution from their combustion products.

Nuclear reactors of the breeder type hold great promise as the solution to these problems.

STATE OF THE LABORATORY, 1970

Alvin M. Weinberg

From *Oak Ridge National Laboratory Review*, Winter 1971, p. 1.

We in the nuclear community have been comfortable in the belief that our work—providing a great new source of energy—is an unmitigated and obvious good. It therefore comes as a perplexing shock to realize that the nuclear community is confronted with what seems to be a crisis of public confidence. Opposition to nuclear energy, which was first expressed publicly seven years ago, by David Lilienthal, has mushroomed. Fanned by well-meaning, but in my opinion poorly informed, scientific polemicists, articulate, though not large, segments of the public are casting doubt on aspects of nuclear energy that we had long since taken for granted. Where we insist nuclear energy is clean, our critics claim it is dirty. Where we insist nuclear energy is safe, our critics claim it is unsafe. Where we insist it is needed for our ultimate survival, our critics say it is unnecessary. As one professional ecologist from Amherst College said in the *American Scientist* (p. 618, November–December 1970), ". . . the technological solutions . . . [based on nuclear breeders] to sustain a world population of 20 billion people are far more immoral than were the decisions to use the A-bomb and H-bomb."

Developments at Oak Ridge, and similar developments elsewhere, justify our belief that in the future it will be possible at reasonable cost to reduce radioactive effluents from chemical processing plants essentially to zero. Thus parts of the nuclear cycle that typically have been responsible for most of the low-level release to the environment can now be completely cleaned up; I would expect that these new techniques will be incorporated into all radiochemical plants.

Another major development in the control of routine nuclear wastes is the decision to proceed with Oak Ridge's salt mine repository in Lyons, Kansas. The Atomic Energy Commission has allocated $25 million for this purpose, thus culminating some 13 years of work and planning at ORNL aimed at sequestering radioactive wastes, finally and forever.

Our decision to go to salt for permanent high-level disposal is one of the most far-reaching decisions we, or for that matter any technologists, have ever made. These wastes will be hazardous for up to a million years. We must therefore be as certain as one can possibly be of anything that the wastes, once sequestered in the salt, can under no conceivable circumstances come in contact with the biosphere. What gives us such assurance about waste disposal in salt?

The primary reason is that the beds of salt, simply because they are still there, could not, since the Permian period (250 million years ago), have been in contact with circulating groundwater. The beds are in a geologically stable region, and have shown no signs of earthquakes. The salt is 1000 feet below the surface, beyond the extent of all previous continental ice sheets. This all but precludes the possibility of disinterment of the wastes during their hazardous lifetime by natural processes operating from the surface.

Finally, there is the question of shipment of radioactive fuel. If, by the year 2000 we have 940,000 megawatts of nuclear power (as is predicted by the AEC) of which two-thirds are liquid metal fast breeder reactors, then there will be 7,000 to 12,000 annual shipments of fuel from reactors to chemical plants, with an average of 60 to 100 loaded casks in transit at all times. Projected shipments might contain 1.5 tons of core fuel with a decay time of 30 days, 300 kilowatts of thermal power, and radioactivity of 75 megacuries. By comparison, present casks from light-water reactors may produce 30 kilowatts and contain 7 megacuries.

Obviously, design of a completely reliable shipping cask for such a radioactive load is a formidable task. As presently conceived, the heat would be transferred to air by liquid metal or molten salt; and the cask would be provided with rugged shields and seals that are resistant to deformation that might be caused by a train wreck.

The shipping problem looms as a difficult one, and it may be that we shall have to change our basic strategy. We may decide to cool fuel in place for 360 days before shipping; this reduces the heat load sixfold, and increases the cost of power by only about 0.2 mill per kilowatt-hour. Or a solution, which I personally prefer, is to cluster fast breeder reactors in nuclear power parks which have their own on-site reprocessing. We estimate that it would cost only 0.1 mill per kilowatt-hour more to reprocess in a nuclear park serving 10,000 electrical megawatts than to ship to a central plant serving 50,000 megawatts; this of course does not count the added cost of power transmission and pollution abatement. But the logic of this trend seems strong, and has led to establishment of Hanford as a nuclear power park.

Of all the components of a PWR, the pressure vessel is probably the most important. If the pressure vessel, as well as the rest of the primary system, maintains its integrity at all times, no incident can present a serious hazard to the public. To reassure ourselves as to the incredibility of a catastrophic pressure vessel failure, Witts's colleagues are devising ways of judging the integrity and safety of very large steel sections by extrapolation from experiments performed on small specimens. The experiments are difficult and expensive; yet the conclusion one must draw is that nuclear pressure vessels are designed very conservatively, and their failure is indeed all but incredible.

If one stands back and tries to assess whether an uncontained nuclear accident is absolutely impossible, one would have to say, no, it is not absolutely impossible; there is some extremely small possibility of an accident. Critics of nuclear power often point to the 1957 AEC study (WASH-740, "Theoretical Possibilities and Consequences of Major Accidents in Large Nuclear Plants," March 1957) which predicts that 3,400 people might be killed and $7 billion worth of property might be damaged if a 200-electrical-megawatt reactor dispersed its fission products. But the critics do not mention that WASH-740 gave no credible scenario leading to the accident it postulates. Moreover, absolutely no credit is taken for realistic mechanisms, some of which I have described, that would greatly reduce the driving force of the accident and reduce the airborne concentration of fission products. Thus in trying to weight this risk one must first point out that the probability of an uncontained accident is extremely small; and second that, because there are many factors that would mitigate the situation, even an uncontained accident would be considerably less serious than suggested in WASH-740.

Some point to fusion power as the clean energy source. Fusion still rides on the wave of optimism that has surged in the last two years, and certainly the Oak Ridge program has taken on a new excitement with major experiments like ORMAK and IMP about to be turned on, and with the new microwave-heated bumpy torus in the offing. This very optimism has kept Oak Ridge in the forefront in the matter of engineering the fusion reactors, and here it is that the safety and environmental aspects have been given further scrutiny.

The greatest hazard in a fusion reactor lies in its huge tritium inventory—typically a hundred million curies. Tritium has a half-life of 12 years, and, to prevent atmospheric buildup, the leakage rates have to be kept below about one-millionth of the inventory per day. Can this be done? I think it can; although there is the same problem of diffusion through hot metal as we have in MSR, nevertheless this is the kind of problem that Oak Ridge has the strength to resolve. What then about catastrophic release? A paper by Herman Postma gives comparative figures as between tritium release from a fusion reactor and iodine-131 release from a fission reactor of equal electrical output. Not only are there four times as many curies of iodine as of tritium; but, because tritium is so much less noxious to biological systems, we end up with fusion presenting at least ten thousand times less potential hazard than fission.

In weighing benefits against risks in the large-scale use of power reactors, we must place in the balance those benefits that directly flow from the use of nuclear power rather than the indirect ones, such as isotopes and basic research. The direct benefits are both long range and short range. Even some of the aggressive critics grant that nuclear power in the long run will be a great boon, that mankind must eventually have an alternative to fossil fuel simply to survive. But they ask, Why should we proceed with nuclear power at our present pace?

The simplest answer is that we need more energy now from every source; and nuclear power is cheaper and is less damaging to the environment than is fossil-fueled power. Only with respect to heat rejected from water reactors does nuclear power do worse than fossil. Chemical effluents from nuclear plants are essentially nil. Even as regards radioactive effluents, a fossil-fueled plant emits more biologically damaging radioactivity than a pressurized water reactor. Of course the effluents from presently designed radiochemical plants more than make up for this; but, as I have said, near-zero-release chemical plants are feasible.

There is another compelling argument for nuclear power in addition to its potentially negligible impact on the environment and its availability at a time of extreme shortage of energy. This is the role that nuclear energy might play in helping to resolve two of the world's most urgent problems—hunger in India and refugees in the Middle East. In dealing with both of these matters, ORNL has been heavily involved.

As for the Middle East, we have completed our studies implementing the Baker Resolution and have forwarded them to the State Department. Is it too fantastic to believe that nuclear power and desalting may present new options to the warring parties and therefore help the cause of peace in the Middle East?

The Indo-Gangetic Plain Project has been submitted by the Bhabha Atomic Research Centre to the Indian Government for approval. The plan, as developed by our Indian colleagues in Trombay, calls for 25,000 tube wells energized by two nuclear reactors. The entire project could be completed by around 1980. When completed, it would produce an additional eight million tons of grain per year.

And finally, I would hope that each of us will rethink the terms of the risk-benefit balance that is being struck by nuclear energy: the possibility of forestalling Malthusian catastrophe, but with a means that poses an extremely small, but nonzero, risk. To those of us who have lived and worked at ORNL almost all our lives, and who see the risk in its proper proportion, this seems like an absurd imbalance in favor of the larger benefit. And indeed, herein lies the crux of the matter. For the risk that I believe must be balanced against the benefit of abundant nuclear energy is not the far-fetched and misplaced concern over low-level radiation or even the extremely unlikely catastrophic nuclear accident. Rather it is the catastrophe that will surely beset the many billions who come after us unless they have an adequate source of energy. Insofar as this generation has a responsibility to future generations, it has the responsibility to develop this new technology: to explore its possibilities, to ferret out its shortcomings, to correct them, and thus to present the future with the means for its survival. This we are now doing at Oak Ridge, and this is what we, as fully responsible scientists and human beings, must continue to do until the full potential of nuclear energy is achieved.

Two editorials by Phillip H. Abelson, Editor, appearing in *Science* **169**, Sept. 25 and Dec. 11, 1970. (Complete text)

SCARCITY OF ENERGY

The United States is now faced with serious short-term and long-term problems in satisfying its needs for energy. In the short-term, there is a scarcity of fuels that meet antipollution regulations; in the long-term, we are faced with depletion of our petroleum and natural gas reserves.

The major air pollutant from stationary sources, SO_2, comes largely from thermal electric power plants. About 57 percent of the fuel for such plants is coal that typically contains 2 to 3 percent sulfur. To diminish air pollution, a number of cities, including New York, have adopted regulations that require in effect that fuels have no more than 1 percent sulfur. The result has been a curtailment of the use of coal in such cities, for only limited amounts of coal with 1 percent sulfur or less are available.

In an effort to comply with the regulations, many utilities have switched to fuel oil. This year the demand for residual fuel oil has already risen sharply but supplies have not increased correspondingly. The United States makes little residual fuel oil. More than 90 percent of the needs of northeastern United States are derived from foreign sources.

Faced with a shortage of oil, some utilities have attempted to turn to natural gas as an alternative. They have found that large supplies of this fuel are not available.

Somehow we will muddle through this present shortage—if necessary, by relaxing somewhat the antipollution regulations. However, the long-term energy problem will require more substantive actions. Not only is the United States depleting its reserves of petroleum and natural gas, but it is not moving decisively to fill the gap. About 74 percent of our total energy requirements are met by oil and natural gas. Importing our total supply of these products would at present cost us about $20 billion a year. We cannot afford such an adverse contribution to the balance of payments.

Some intermediate-term relief could be obtained by granting higher prices to gas producers and by opening additional areas of the continental shelf.

The longer-term solutions to our energy problems involve becoming more prudent in the use of energy. The solutions also demand the skillful employment of coal and atomic energy. In principle, all our energy needs could be met for a long time with coal. This raw material could be processed to yield sulfur-free fuel, liquid hydrocarbons, and methane. In practice, however, the development of the use of coal is limping along and is underfinanced. A few hundred million dollars a year devoted to research, development, and demonstration plants could be the most valuable expenditure the government could make.

COSTS VERSUS BENEFITS OF INCREASED ELECTRIC POWER

Typical estimates of future demand for electric power in the United States assume a continuation of the previous rate of growth; power consumption eight times that of the present is projected for the year 2000. Little attention is devoted to the anatomy of the future demand. It is pointed out that population is growing, the gross national product is expanding, and energy demands are expected to increase. However, it is physically impossible for exponential growth to continue indefinitely. Already it is apparent that the generation and distribution of electricity entails some damage to the environment. If conventional fuels are employed, the increased demands on them will speed exhaustion of oil and gas, and the use of large quantities of coal is likely to despoil large areas. Nuclear power carries with it many risks. Thus the utilities can expect to face continuing opposition in their efforts to expand power generation. The outcome of the battle is likely to rest on a balancing of social costs versus benefits to the consumer.

Much of the electric power goes to industry and to commercial use. However, the public is most immediately affected by that part going to individual consumers, and the electorate is likely to base many of its attitudes on personal experience.

If private consumers were to increase their use of power by a factor of 8 by the year 2000, where would the demand come from? Only a small fraction of the increase would come from population growth. There continues to be a proliferation of electrical gadgetry, but power consumption by most of these devices is trivial. For example, an electric razor consumes only a kilowatt hour per year, which is less than an air-conditioned house uses in an hour. In general, the devices that are used intermittently consume only modest amounts annually. Major items and their approximate typical annual consumption in kilowatt hours are color television, 500; lighting, 600; electric range, 1200; frost-free refrigerator-freezer, 1700; freezer, 1700; water heater, 3500; air conditioning, 5000; home heating, 20,000.

The more affluent segments of society already have about all the television sets, lighting, and cooling that they can use. Future expansion in public power consumption is dependent on an increased standard of living by the less affluent and on widespread adoption of electricity for home heating. At present only about 3.5 million homes are heated electrically; the major potential market is in home heating.

Utilities are responding to the public's concern about pollution by extolling the virtues of clean heat. They soft-pedal the fact that the pollution problem is merely transferred elsewhere. However, it is technically much more feasible to eliminate pollution at a few major emitters than in millions of individual homes. Another consideration is the thermodynamic inefficiency introduced when electrical energy is dissipated resistively. However, if heat pumps were utilized at the homes, the overall efficiency would be acceptable.

The era of unquestioned exponential growth in electric power has come to an end. The future course of expansion will be determined by the public's estimate of costs versus benefits.

PLUTONIUM: REACTOR PROLIFERATION THREATENS A NUCLEAR BLACK MARKET

Deborah Shapley

From *Science* **172**, 143 (1971).

The commonly accepted solution to national energy needs is sprinkling the landscape with plutonium-fueled nuclear power reactors. Environmental defenders have argued loudly that these pose health risks, but a less well known threat would arise from a vastly increased traffic in plutonium fuels. It has been predicted that these commercial reactors will cause civilian plutonium stores in the United States to rise from the present 600 to 720,000 kilograms or more in the next 29 years. Even Atomic Energy Commission (AEC) commissioners who are promoting the reactors admit that it is "likely" that this vast traffic will spring a leak onto a worldwide black market. Since plutonium is the stuff from which Nagasaki-type atomic bombs are made, such a black market could put the 5 kilograms of plutonium it takes to make an atomic bomb into the hands of anyone willing to pay.

But the prevention of just this sort of diversion has been a cornerstone of American diplomatic and strategic policy for two decades. It was the prime aim behind the nuclear Non-Proliferation Treaty (NPT) signed in 1968 by 63 nations and ratified last March by 48.

With plutonium as the product as well as the fuel for all these [future reactors], an enormous commercial trade is inevitable, and, in fact, has already started with industry stockpiling of PWR-produced plutonium.

In perhaps the frankest speech on record on safeguarding these materials, AEC Commissioner Clarence E. Larson told a Los Alamos safeguards conference in 1969 about the "likely" nuclear black market of the future.

"Once special nuclear material is successfully stolen in small and possibly economically acceptable quantities, a supply-stimulated market for such illicit materials is bound to develop. And such a market can surely be expected to grow once a source of supply has been identified. As the market grows, the number and size of thefts can be expected to grow with it, and I fear such growth would be extremely rapid once it begins. Such a theft would quickly lead to serious economic burdens to the industry and a threat to national security."

Larson also quoted an "unavoidable" loss rate by industry of 1 to 2 percent: "We in the industry recognize this to be a fact." And later, he admitted, "from a practical point of view, we may never solve all the problems" of safeguarding materials.

Practically everyone in and out of AEC agrees that there is a significant security threat associated with large-scale reactor use. At the Los Alamos conference, Don Povejsil, of Westinghouse's Nuclear Fuel Division, termed the plutonium-guarding problem the "dominant" one. More recently, Ralph F. Lumb, chairman of a 1967 AEC advisory committee on safeguarding materials, specifically referred to the future plutonium problem "like any other business . . . the more you have of something, the more you're going to lose. It's something in the nature of risks."

At present, plutonium sells for about $10,000 per kilogram. It is thus five times as costly as heroin and ten times as expensive as gold. What its value would be on an illegal market is anybody's guess.

It takes only a very small bit of plutonium—about 5 kilograms or $50,000 worth—to make a bomb the size of the weapon that destroyed Nagasaki in August 1945. The technology and hardware are available —many sources recommend the *World Book* as a good text on atomic bomb-building.* Finally unlike the uranium now used in reactors, plutonium is relatively easily processed into weapons-ready condition. Hence the only real obstacle now stopping anyone from building a perfectly good bomb is the present scarcity of the materials and the tight

---

* In the author's opinion this is an overstatement, but the required technical talent is presumably available.

security kept by the five nuclear powers over their uranium and plutonium.

To date, there have been a few reported losses of strategic material, although only one is said publicly to have involved an attempted theft. In late 1969, the experimental SEFOR reactor in Strickler, Arkansas, was found deficient in "a few kilograms" of plutonium. A Nuclear Materials Enrichment Corporation plant in Apollo, Pennsylvania, discovered that 6 percent of its materials had gone unaccounted for over a 6-year period. In both cases the AEC attributed the losses to normal processing. In Bradwell, England, however, two reactor plant workers dropped 20 fuel rods over the plant fence and left them, apparently to be picked up. The theft was intercepted, however.

The countermeasures to this scenario adopted by AEC's Office of Safeguards and Materials Management (OSMM) is to keep continuous audits on the materials themselves at each stage of processing and use. But such audits can never keep track of 100 percent of the material: inevitably, some is lost in normal chemical reactions, some sticks in pipes, in vents, to workers' uniforms and gloves, and some falls in among the scrap. The purpose of the audits is to determine what loss rates are tolerable so that any unusual losses will be known instantly. In that case, AEC can require a facility to shut down and clean itself inside out to find the lost material. According to Delmar L. Crowson, head of OSMM, audits for plutonium-239 loss average within $+0.18$ to 0.51 percent, with 0.2 percent "not unusual."

However, scientists working for AEC on this problem say that the above figures do not reflect the margins of uncertainty involved—which sometimes run as high as 1 percent or more. Current calculations, they say, are based on estimates, for example, of how much strategic material might be in a scrap heap—not on precise, actual measurement.

AEC is now working to revise its guidelines for industry so as much guesswork as possible will be eliminated. Thus, when the amount of material involved begins to climb, and IAEA begins its formal inspections in March 1972, AEC will have a more realistic tab on the materials.

But critics of AEC safeguards find the holdup or hijacking scenario more likely—and less well guarded against—than smuggling. Crowson told *Science* that the most likely point for materials theft is the fuel reprocessing plant. But critics, including Dr. Theodore Taylor, former safeguards consultant to the AEC, believe that loading, shipping, and transfer processes are most vulnerable. Dr. Taylor believes there is a good chance that the planes which carry the materials by commercial air freight could be hi-jacked.

Shipping: The Weakest Link

One of the anachronisms of AEC policy is that strategic nuclear materials which are to be used for military purposes are shipped under military rules. But, if the same materials are to be used for civilian purposes —although they too could fuel a bomb—they are usually shipped, in the words of Crowson, "like a special delivery letter."

Part of the AEC's mandate is to promote private industry. In this case, it gives its business to commercial carriers. Sometimes the carriers, particularly railroads, have refused to ship it on the grounds it was too dangerous. But the question of tighter controls on the truckers and freight companies is a touchy one. At the Los Alamos meeting which Commissioner Larsen addressed, one trucking consultant retorted that making security checks on shipping personnel for example, was "retrogression, not progression. I was under the impression that the whole program of the AEC was to turn things over to industry. . . . I resent the implication that only the government is capable of doing anything correctly."

Sam Edlow, a transport consultant who has arranged shipments of fissionable materials internationally, said, "The carrier agrees to deliver a specific shipment between two specific points, at a published freight rate within a reasonable period of time. . . . He doesn't promise you that he is going to follow your instructions at the transfer point of connecting carriers. . . . He won't guarantee to do it in a specific period of time. . . . If the Commission is serious—really serious—about establishing a real set of safeguards within the transportation cycle, the answer is probably (in addition to making the shipper act in a professional way) regulation of the transportation industry itself, because only in that way can authority be expressed on the industry."

But at the same time, the truckers agreed that organized crime could easily obtain nuclear materials if it wanted. "Anything that organized crime wants to lay its hands on, while it's in the transportation cycle, it's going to get."

AEC hired Wright, Long & Co. to make a study of the threat of hi-jacking by the Mafia and other organized groups. The study itself is classified, but Carmine Bellino of Wright, Long & Co. told the Los Alamos meeting that "on a list of 735 so-called Mafia members, 12 are or were owners of trucking firms, two are truck drivers, and at least nine were union officials." While interviews with police chiefs had revealed that the Mafia appeared more interested in cigarettes and television sets than in uranium and plutonium, he added: "It is possible, however, they would add, that some foreign tyrant might offer a deal of some kind to any racketeer who would divert enriched uranium or plutonium. . . . In such a situation a truck carrying uranium or plutonium could be easily hi-jacked or the theft could occur at warehouse or dockside." If such a threat exists already, the situation can only grow when, in 1974 and 1975, much greater quantities of plutonium will be shipped and stored.

Information Access

Although the black market problem is recognized by many at AEC as a grave question facing the nuclear reactor program and its plans for the future, to date there has apparently been little effort to make it public.

Naturally, details of losses, such as those at the Arkansas SEFOR reactor, are kept quiet for purposes of security. Congressional hearings dealing with security measures are also, generally, closed to the public. But the basic problem—the security aspect of AEC's future reactor program—is scattered through technical reports and documents, and it has rarely reached the public eye in coherent form.

When the head of the safeguards division first outlined the work of his group to *Science*, the problem of tracking great quantities of plutonium on a vast scale in the future was not even mentioned. (Later, when asked about the problem, he discussed it at some length and described his division's studies on the subject. The studies, however, are not public.)

The hazy line between withholding information on the problem and obfuscating it stretches into Commissioner Larsen's speech on safeguards. There he repeated what many others in the field have said: that industry losses of strategic materials can run about 1 percent of the total. But when Larsen's speech was published with the other symposium proceedings, this admission had been reworded to say that "small process losses are unavoidable." The attitude at AEC seems to be to avoid telling the public much about this problem until the agency thinks it has a solution well in hand.

REMARKS BY CONGRESSMAN CRAIG HOSMER

Before the Eleventh Annual Meeting
of the Institute of Nuclear Materials Management
Gatlinburg, Tennessee, May 25, 1970

Let us briefly trace the course of our United States national approach to special nuclear materials accountability.

In Manhattan District days and up to just sixteen years ago the materials were kept tightly in the possession of the Government. How much oralloy and how much plutonium are unaccounted for during this period is not for me to say. Actually, I don't know and I doubt if anyone else could come up with much more than an educated guess. But since the amount of these materials was relatively small during this period, I believe we can safely say that losses likewise were relatively small.

This initial period ended in 1954 with the extensive revision of the Atomic Energy Act. For the first time special nuclear material was made available legally to private persons for peaceful purposes. The Commission established procedures and criteria for the issuance of licenses to receive, use and transfer SNM [Special Nuclear Materials]. It proceeded on the general assumption that the financial responsibility of licensees for loss and damage, together with the severe criminal penalties written into the Act, would result in safeguards being initiated by licensees to protect their pocketbooks which, in the process, also would serve adequately to protect the material.

Two years later, in 1956, came the United States Atoms for Peace program with its authority for the export of SNM to cooperating nations under bilateral agreements giving the United States ample rights to implement safeguards inspections and control measures in the receiving country. This early American precedent continues in effect today not only in our current bilaterals, but also in almost identical terms in the International Atomic Energy Agency's statute on the subject. Undoubtedly it will carry through into the safeguards agreements nonnuclear powers will be making with IAEA in compliance with the Nonproliferation Treaty.

The cumulative, worldwide production of plutonium in nuclear power reactors has been estimated to be approximately 125,000 kilograms by 1975 and almost three million kilograms by 1985. That is enough, someone has told me, to make 15,000 or more fission bombs at the first date and hundreds of thousands by the second.

These kinds of estimates are nothing new and nothing secret. Because of the general public interest in their troublesome implications the United States arrived at another milestone in its domestic safeguards program in 1967. At that time the Commission took two significant steps in the interest of strengthening its ability to meet the growing need for practical and effective safeguards measures:

It established the Division of Nuclear Materials Safeguards under the wings of Harold Price, Director of Regulation, with the responsibility for administering the safeguards program with respect to Commission licensees, and

For the purpose of implementing safeguards with respect to license exempt contractors and AEC operated facilities, to conduct a more aggressive safeguards R & D program, and, as the center for developing both domestic and international safeguards policies, AEC established the Office of Safeguards and Materials Management.

At your Seventh Annual Meeting in 1966, the then Executive Director of the Joint Committee on Atomic Energy, John Conway, told you of

(*context*) Numec Corporation's unaccounted for rate of 6 percent on highly enriched uranium during fuel fabrication and scrap recovery processes under United States contracts. In excess of 100 kilograms of this precious material were unaccounted for by Numec over a period of years. I know of no subsequent case of this magnitude. (*context*)

Now, before some sensation hunting writer takes the foregoing out of *context* and tries to use it to hit the atom in this country, the government, the AEC and the JCAE over the head with it, let me warn him to put it right back in context. I didn't say 100 kilograms disappeared, or was stolen, or was lost or is now in the hands of country X or the Mafia. I said that Numec's books and records failed to show uranium out of its plant *vice* uranium in by a discrepancy of 100 kilograms. The complaint is not necessarily regarding *actually* disappearing enriched uranium. It is about bad process control and accountability practices which, over a period of years, got the record books drastically out of balance and cost

Numec a lot of money. We know for certain that more uranium did not come out of Numec than went in—and we can say that less went out than went in.

And, until the AEC gets around to carrying out the injunction to define what normal losses are at various stages of the nuclear process, even properly kept books are not going to tell us as much as we ought to know. Numec's job was a difficult one. Maybe a 6 percent loss during the process it was carrying on should be regarded as normal. I doubt it. But we'll never know until AEC does establish those norms.

Undoubtedly accountability has improved since 1966 and I don't want to sound gloomy. However, lest the safeguards fraternity become complacent, let me cite to you three recent examples of what are euphemistically called "misroutings" during SNM shipments:

In March, 1969, a container of highly enriched UF_6 was scheduled to go from Portsmouth, Ohio, to Hematite, Missouri. It didn't get there. The AEC, the FBI, the airline, the police, and untold numbers of individuals searched in vain for the shipment which was dispatched on March 5th. Finally, on the fourteenth, it was located in Boston.

Also in March, 1969, highly enriched uranium was booked for departure from New York's Kennedy International Airport on the 11th for delivery to Frankfurt, Germany, on the afternoon of the 12th. The material did not arrive. Five days later, on March 17th, it finally turned up in London where it apparently had been offloaded in error, and,

Only last month a drum of waste containing a small amount of 70 percent enriched uranium was consigned for delivery from one firm to another in the same California city. It was, instead, carelessly sent to Tiajuana, Mexico. The report on this matter was imaginatively entitled "Inadvertent export of special nuclear materials."

In these three cases all indications point to slipshod, slapdash handling by shippers. Nobody got hurt. No financial loss ensued. No material went unrecovered. But these happenings dramatically point up a need for more effective safeguards during SNM shipments. I am pleased to say that the AEC has responded with a considerable tightening up of its shipping regulations. The Commission and all other safeguards authorities, national and international, have a constant duty to improve their monitoring capabilities at any and all points where accidental or deliberate diversions might take place. It is important to remember that if some of the stuff can get lost through carelessness, an awful lot more of it could disappear if some people put their minds to stealing it for illicit profit.

Earlier this year the Attorney General of the United States cited the Kennedy Airport cargo handling apparatus as being under the control of organized crime. The same can be said of many other key transportation elements of this country too. When and if SNM ever becomes an article of illicit commerce, the transportation element of the nuclear fuel cycle will become most vulnerable to diversions. We'd better be cinching up in this area all along the way.

And, I should add, all around the world, too. It seems reasonable to assume that if discrepancies such as I have mentioned can occur in The United States, which has had more than a quarter-of-a-century's experience handling special nuclear materials, then there may be many more places here and there which need even more accountability and safeguards work done.

That is where, of course, the International Atomic Energy Agency comes into the picture. Even before the NPT spelled out specific additional safeguards duties for IAEA, the United States strongly urged that IAEA's safeguards role be enlarged. Other of the Lumb Panel's additional recommendations were to spotlight the IAEA *as the operator of a Universal Safeguards system* and to *establish international safeguards inspector training schools.* The AEC has, in fact, established such a school at the Argonne National Laboratory to which international attendance always is invited.

Actually, the IAEA inspections to which a number of United States nuclear facilities are being voluntarily subjected are discreetly used to forward training of international inspection personnel in the techniques of their trade. This is a bonus dividend on top of the stated purpose of these voluntary inspections, namely, to set to rest by our own show of confidence in IAEA's integrity the non-nuclear nations' claims that the inspectors will steal their trade secrets.

Whither the IAEA?

Now I would like to examine the role of international and national safeguards systems in a somewhat broader context with the idea of approximating a real-world picture of what they ought to look like and what they ought to accomplish.

Aside from the possibility of diversion to military or other mischievous uses, there are several reasons why special nuclear materials have to be managed skilfully. One is economic. The stuff is valuable. Like gold, platinum, diamonds and rubies, there are dollar penalties for failure to keep track of it. Another reason is the possible danger to public health and safety if SNM, especially plutonium which is highly toxic as well as radioactively dangerous, is allowed to be spread around carelessly.

Like other police organizations, the Agency safeguards setup is going to do its job well. It will go about it using a combination of safeguards *modus operandi* including,

The "Chastity Belt" approach involving such things as seals on reactors and other SNM containers.

The "Slaughterhouse" approach where figuratively international inspectors wander around nuclear facilities stamping Good Housekeeping seals on SNM accountability practices. And,

The "Black Box" approach which incorporates tamper-proof non-destructive testing apparatus at strategic links in the nuclear fuel chain and elsewhere.

Actually within a very short time I believe the IAEA will be doing everything that can be reasonably expected of it. But just like any other police force that doesn't mean it will be able to stop all crime. There is some finite possibility that some one or more NPT signers will cheat, get away with it, and obtain a surreptitious Nuclear Club membership card. There is a probability that one or more NPT signers will simply denounce the Treaty and go down the nuclear road openly. Then, of course, there are always the countries who refused to sign the Treaty in the first place, who have the capability to go nuclear, and might develop a determination to do so.

Actually, where I think the IAEA will do a tremendous job itself, and by its example and pressure encourage the national systems to do a better job, is in the non-national nuclear threat area. Many people, including myself, do not regard as very convincing the Dr. Goldfinger scenario where James Bond thwarts police holding Miami hostage for a zillion dollar ransom under threat of blowing it up with a stolen H-bomb. Stealing a 1000 pound top secret bomb isn't exactly easy.

But when you think not in terms of stealing whole bombs, but of diverting very small amounts of SNM at a time and of the possibility of a profitable Black Market developing, you get on more credible ground. Black Markets already exist for all kinds of "hot" goods. They are quite flexible in taking on new product lines. If a SNM Black Market develops, the sales price to some country, individual or organization desperately wanting to make nuclear explosives has been estimated as high as $100,000 per kilogram.

A gram is 1/1000th of a kilogram and 1/1000th of $100,000 is $1,000. Liberating a half gram of plutonium at a time from the local fast breeder reactor fuel element factory might be so small an amount as to be relatively undetectable even by the best black boxes and the sharpest eyed inspectors. Kimberley has tried to stop employees from stealing its diamonds for almost a hundred years and hasn't entirely succeeded yet. Even if the stolen material must be sold through a fence at a knock-down price, some employees of the factory may see the risk-to-benefit ratio of this kind of extracurricular activity as favorable.

To reduce both national and non-national threats of this kind I have two specific suggestions which are equally applicable to the international and the various national safeguards systems:

First, SNM safeguards organization and personnel should develop intimate ties with all existing police type organizations to the end that all of the latters' widespread apparatus and resources continuously and effectively will augment the safeguards system.

Second, that the IAEA, the major nations individually, establish rewards for information leading to the arrest and conviction of anyone illegally diverting, holding and using SNM. A "no questions asked" bounty system also might be established for return of unaccounted for material to proper authorities.

I believe we should recognize that rewards for information and stolen items have been used by security and law enforcement officials since the beginning of mankind. Informants are the backbone of any security apparatus.

STATEMENT OF SENATOR MIKE GRAVEL

On introduction of a bill to repeal major provisions of the
Price-Anderson Act regarding liability for nuclear accidents
May 13, 1971

Mr. President, I am introducing a bill which would repeal the main
provisions of the Price-Anderson Act (Public Law 85-256), for I believe
anyone who uses nuclear material should accept the financial liability
for damage which that nuclear activity may cause to the public. My
belief is based on the simple observation that, in our economic system,
public liability is the principal restraint on reckless activity. I fundamentally question the wisdom of the Price-Anderson Act, which, "in
order . . . to encourage the development of the atomic energy industry . . ." stipulates that "The United States may make funds available
for a portion of the damages suffered by the public from nuclear inci-
dents, and may limit the liability of those persons liable for such losses."

The word "incident" replaces the word accident, but a single man-
made nuclear incident causing more than five hundred million dollars
worth of damage is an *accident* in anyone's language. It's a man-made
catastrophe.

One important provision of the Price-Anderson Act, which added
Section 170 to the Atomic Energy Act of 1954, sets the limit for public
liability at $560 million per nuclear accident, regardless of the real size
of the damage, which could exceed $7 billion per accident according
to the Atomic Energy Commission (July 1970).

The other most important provision of the Price-Anderson Act
(see AEC regulation 10 CFR 140.11 which implements it) is that about
80 percent of the $560 million must be paid to the injured parties by the
American taxpayers, not by the AEC license-holder.

My bill addresses both of those key provisions: the proposed
section 170a. removes the limit on the liability, and it removes all of
the taxpayer's liability for the nuclear hazards created by license-holders
such as electric utilities and universities. As proposed, it stipulates that
whosoever chooses to put the public at risk with his nuclear activity
must demonstrate, as a condition of receiving and retaining his license,
that he has and maintains financial protection to cover public liability
claims.

The proposed sections 170a. and 170i. retain the sound concepts presently embodied in section 170 that license-holders and contractors waive the issue of charitable or governmental immunity and waive any immunity from public liability conferred by federal or state law, waive the issue of fault, and waive the statute of limitations if suit is instituted within three years from the date on which the injured person could first have known of his injury (e.g., radiation-induced cancer, which may not become apparent for 10–30 years).

In fact, these very provisions in the present version of section 170 reflect one of the reasons that Price-Anderson should not simply be repealed. Complete repeal might leave no one at all liable even for power plant accidents, since a publicly regulated utility might claim government immunity for performing a public service, or might be protected against suits by the inability of the public to prove negligence or fault from the radioactive and unapproachable debris of the former power plant. Furthermore, simple repeal would undo the progress embodied in the sections of 170 which apply to government-sponsored nuclear activities—such as underground nuclear bomb-tests. Namely, in section 170, the AEC circumvents the principle of "sovereign immunity" by agreeing to indemnify its contractors for public damages up to $500 million. My proposed bill would retain the principle of AEC financial responsibility for what it causes to be done, but would simply remove the artificial limit on the amount of its liability.

Perhaps because of some of the provisions described above, the Price-Anderson Act claimed in its opening paragraph not only to encourage the atomic energy industry, but also to protect the public. By some illogic, it was construed to be public protection when the public both suffers an extraordinary man-made disaster and then has to pay for most of the damage.

Nevertheless, Congress did endorse this kind of reasoning when it passed the Price-Anderson Act by voice-vote in 1957. Repeal of the Price-Anderson Act will obviously require the government to offer relief measures for present license-holders who prefer to terminate nuclear operations rather than accept the financial responsibility.

The Price-Anderson Act, which was passed explicitly "to encourage the development of the atomic energy industry," may simply be obsolete in 1971. Do we really need to stimulate the atomic energy industry, in particular, nuclear electricity? We now know, for instance, that solar energy techniques can make it possible to power civilization with sunlight instead of with nuclear fuel. A "sunshine economy" appears as feasible as a "radioactive economy," and far less dangerous.

The AEC expects to license about 1,000 nuclear power plants in the next 29 years.

Since their non-disposable radioactive˙ by-products clearly have the power to poison the entire planet permanently and completely, the AEC is forced to presume that man will be able to confine the radioactive poison he is creating with 99.999 percent success for hundreds—no, thousands—of years to come.

All that stands between nuclear electricity and irrevocable radioactive poisoning of the planet Earth is near-infallibility in the nuclear electricity industry.

Therefore, it is disconcerting in the extreme to realize that the nuclear electricity industry is rapidly expanding under a law—the Price-Anderson Act—which not only acknowledges that giant nuclear accidents can happen, but then proceeds to remove the very restraint which normally operates to prevent reckless activities, namely full liability for public damages.

When the Price-Anderson Act was first passed and then renewed, utility representatives testified that they would build no nuclear plants if they had to stand fully liable for accidents.

As millions of magazine-readers know, the electric utilities now vigorously deny the basic premise of the Price-Anderson Act—that giant nuclear accidents can happen. In a two-page advertisement called, "Go play in the nuclear power park," which appeared in *Newsweek* September 21, 1970 and in *Look* on October 6, 1970, about seventy investor-owned power companies made the following claim: "Before the go-ahead is ever given to build a nuclear power plant, the Atomic Energy Commission requires that the potential owner adhere to safety standards that will withstand every conceivable emergency. . . ."

The utilities sound very sure indeed now. Let us briefly consider the safety claims of Portland General Electric, which recently received a construction permit for its Trojan nuclear power plant.

With reference to the Trojan approval letter written by the AEC's Advisory Committee on reactor safeguards, PGE claimed in writing just a few weeks ago that "the best technical minds in the nation (the ACRS) can not hypothesize any accident or operating condition where the

(Trojan) plant would be unsafe except these few areas which have been suggested by this Committee." PGE was indicating that, after it made the design changes in "these few areas" requested by the Advisory Committee, a public hazard from the Trojan nuclear plant would be inconceivable, even to the best technical minds in the country.

Surely a company so confident that giant accidents are inconceivable does not need the Price-Anderson Act to protect it from public liability suits.

Explicitly, PGE has stated that even if its proposed Trojan plant were to have an accident, at the worst it would kill no one, possibly injure 6 people, and require the evacuation of 67 people. This claim was based on "the continuous operation of the emergency core cooling system after the initial fission-product release" inside the building, and based on no rupture of the containment.

Asking people to bet their lives on the "emergency core cooling system" is like asking the public to test fly a new airplane. The reason I say that is because no emergency core cooling system has ever been fully tested to see if it will really work when the chips are down. The AEC plans the first large-scale tests in 1975, a year after the Trojan plant and many many others are due to go into operation.

Mini-scale tests of the emergency core cooling system have recently cast new doubt on the effectiveness of that vital safety system (*Nucleonics Week*, May 6, 1971).

Do the directors of PGE continue to stand behind their company's supersafety claims? If so, then we can certainly expect those men to demonstrate their sincerity by supporting my bill. After all, this bill merely makes their company liable for accidents they have claimed will never happen. Likewise, I see every reason for the directors of all the investor-owned power companies which sponsored the advertisement in *Newsweek* and *Look* actively to support my bill, preferably in new coast-to-coast advertisements.

My point is very simple; they have claimed that their nuclear power plants can withstand every conceivable emergency, and it's time for them to put their money where their mouths are. If they won't even risk their dollars on nuclear electricity, why should Americans be forced to risk their lives?

If it is no longer conceivable for nuclear power plants to have the disastrous radioactive releases which prompted the Price-Anderson

Act in the first place, then the utilities no longer need the protection of that act.

If it is possible for catastrophic nuclear accidents to happen, then the public is entitled to hear that message, too, in double-page nationwide advertisements.

The public is entitled to know how great the public injury might be from the worst possible accident in terms of extra cancers, defective babies, and contaminated property. $20 billion? $10 billion? $2 billion? $100 million? Or zero, as PGE claims? As of July 1970, the AEC was standing by its 1957 estimate—$7 billion or worse for the maximum credible accident. The utilities deny it.

Some people have warned me that my bill would kill nuclear electricity simply because no single utility has sufficient assets to cover the possible public liability from a nuclear accident. To consider such a warning, first we must determine how large the possible public injury could really be. Zero, or colossal? My bill would clearly require an inquiry into that very question—one for which the public deserves an unbiased and believable answer instead of totally contradictory claims.

The utilities have everything to gain by such an inquiry, providing their safety claims are true. However, if such an inquiry determines that possible public injury from nuclear power plants is great enough to exceed the assets of licensed utilities, then it is surely time to re-evaluate the wisdom and morality of licensing such machines at all, especially in view of the safe alternatives.

CONTROLLED FUSION RESEARCH AND HIGH-TEMPERATURE PLASMAS

Postscript to an article by R. F. Post

From *Annual Review of Nuclear Science* **20**, 509 (1970).

The Next Ten Years—and Beyond

Prediction is risky and subjective; history can be interpreted in more than one way. Nevertheless I could not conclude this compendium without attempting a little of both.

In the last 11 years we have seen the growth to approaching maturity of a new field of endeavor—high-temperature plasma physics. Brought to sharp focus by the quest for fusion power, the result has, it seems, been to lead us close to the point where scientific feasibility (T and $n\tau$) of fusion will be proved, perhaps within another 5 years.

As to the present worldwide effort, from which the push toward scientific feasibility must begin, it is about 120 million dollars equivalent per year; roughly 50 to 60 percent is in the USSR, 25 percent in the United States, and the rest in Western Europe and Japan. From this present level of effort one would conclude that the first demonstration of scientific feasibility is most likely to come either from the USSR or the United States.

A worldwide level of only 120 million dollars seems a small effort, considering the rewards for success. At the very least the effort devoted to fusion research over the next 5 years should be doubled. At that level, when coupled with the concentration of attention that proof of scientific feasibility should permit, the pace toward a reactor should quicken. I believe that in less than 5 years after the demonstration of scientific feasibility a pilot plant reactor could be running. Given the advanced technological base we can logically expect 10 years from now, I see no reason why fusion plants would not begin to make a substantial contribution to the power industry before 1990. As I said before, whether this goal will be achieved, given proof of scientific feasibility, will depend more on national priorities and on the will to accomplish it than on any other factor. I am in complete disagreement with those who say What's the rush?—I believe fusion power should be avidly sought for, not waited for. Does a permanent solution to mankind's energy needs deserve any less?

The Decision to Use A-Bombs
in World War II

Excerpts from the Franck Report

These excerpts come from a report submitted to the Secretary of War, Henry L.
Stimson, on June 11, 1945, by a committee consisting of three physicists, three
chemists and one biologist, with James Franck as chairman, appointed by the
director of the Chicago laboratory of the "Manhattan Project," Arthur H.
Compton. The report is published essentially in full in the *Bulletin of the Atomic
Scientists* 1 (1 May 1946).

The excerpts from this and the following article constitute about a third to a
half of the two documents. The complete texts may be found not only in the
sources quoted but also in the Appendix of *The Discovery of Nuclear Fission*, by
H. G. Graetzer and D. L. Anderson, Van Nostrand Reinhold Company,
New York (1971). For discussion see *The Atomic Bomb—The Great Decision*,
edited by P. R. Baker: Holt, Rinehart and Winston, New York (1968).

The only reason to treat nuclear power differently from all the other
developments in the field of physics is the possibility of its use as a
means of political pressure in peace and sudden destruction in war. All
present plans for the organization of research, scientific and industrial
development, and publication in the field of nucleonics are conditioned
by the political and military climate in which one expects those plans to
be carried out. Therefore, in making suggestions for the postwar
organization of nucleonics, a discussion of political problems cannot
be avoided. The scientists on this Project do not presume to speak

authoritatively on problems of national and international policy. However, we found ourselves by the force of events, during the last five years, in the position of a small group of citizens cognizant of a grave danger for the safety of this country as well as for the future of all the other nations, of which the rest of mankind is unaware. We therefore feel it our duty to urge that the political problems, arising from the mastering of nuclear power, be recognized in all their gravity, and that appropriate steps be taken for their study and the preparation of necessary decisions.

Scientists have often before been accused of providing new weapons for the mutual destruction of nations, instead of improving their wellbeing. It is undoubtedly true that the discovery of flying, for example, has so far brought much more misery than enjoyment and profit to humanity. However, in the past, scientists could disclaim direct responsibility for the use to which mankind had put their disinterested discoveries. We feel compelled to take a more active stand now because the success which we have achieved in the development of nuclear power is fraught with infinitely greater dangers than were all the inventions of the past. All of us, familiar with the present state of nucleonics, live with the vision before our eyes of sudden destruction visited on our own country, of a Pearl Harbor disaster repeated in thousand-fold magnification in every one of our major cities.

In the past, science has often been able to provide also new methods of protection against new weapons of aggression it made possible, but it cannot promise such efficient protection against the destructive use of nuclear power. This protection can come only from the political organization of the world. Among all the arguments calling for an efficient international organization for peace the existence of nuclear weapons is the most compelling one. In the absence of an international authority which would make all resort to force in international conflicts impossible, nations could still be diverted from a path which must lead to total mutual destruction, by a specific international agreement barring a nuclear armaments race.

One thing is clear: any international agreement on prevention of nuclear armaments must be backed by actual and efficient controls. No paper agreement can be sufficient since neither this or any other nation can stake its whole existence on trust in other nations' signatures. Every attempt to impede the international control agencies would have to be considered equivalent to denunciation of the agreement.

SUMMARY

The development of nuclear power not only constitutes an important addition to the technological and military power of the United States, but also creates grave political and economic problems for the future of this country.

Nuclear bombs cannot possibly remain a "secret weapon" at the exclusive disposal of this country for more than a few years. The scientific facts on which their construction is based are well-known to scientists of other countries. Unless an effective international control of nuclear explosives is instituted, a race for nuclear armaments is certain to ensue following the first revelation of our possession of nuclear weapons to the world. Within ten years other countries may have nuclear bombs, each of which, weighing less than a ton, could destroy an urban area of more than ten square miles. In the war to which such an armaments race is likely to lead, the United States, with its agglomeration of population and industry in comparatively few metropolitan districts, will be at a disadvantage compared to nations whose population and industry are scattered over large areas.

We believe that these considerations make the use of nuclear bombs for an early unannounced attack against Japan inadvisable. If the United States were to be the first to release this new means of indiscriminate destruction upon mankind, she would sacrifice public support throughout the world, precipitate the race for armaments, and prejudice the possibility of reaching an international agreement on the future control of such weapons.

Much more favorable conditions for the eventual achievement of such an agreement could be created if nuclear bombs were first revealed to the world by a demonstration in an appropriately selected uninhabited area.

In case chances for the establishment of an effective international control of nuclear weapons should have to be considered slight at the present time, then not only the use of these weapons against Japan, but even their early demonstration, may be contrary to the interests of this country. A postponement of such a demonstration will have in this case the advantage of delaying the beginning of the nuclear armaments race as long as possible.

If the government should decide in favor of an early demonstration of nuclear weapons, it will then have the possibility of taking into account the public opinion of this country and of the other nations before deciding whether these weapons should be used against Japan. In this way, other nations may assume a share of responsibility for such a fateful decision.

Remarks by Secretary of War Henry L. Stimson

From *Harper's Magazine* **194,** 97–107 (February, 1947)

Hiroshima was bombed on August 6 and Nagasaki on August 9, 1945.

Many accounts have been written about the Japanese surrender. After a prolonged Japanese cabinet session in which the deadlock was broken by the Emperor himself, the offer to surrender was made on August 10...

The two atomic bombs which we had dropped were the only ones we had ready, and our rate of production at the time was very small. Had the war continued until the projected invasion on November 1, additional fire raids of B-29's would have been more destructive of life and property than the very limited number of atomic raids which we could have executed in the same period. But the atomic bomb was more than a weapon of terrible destruction; it was a psychological weapon. In March 1945 our Air Force had launched its first great incendiary raid on the Tokyo area. In this raid more damage was done and more casualties were inflicted than was the case of Hiroshima. Hundreds of bombers took part and hundreds of tons of incendiaries were dropped. Similar successive raids burned out a great part of the urban area of Japan, but the Japanese fought on. On August 6 one B-29 dropped a single atomic bomb on Hiroshima. Three days later a second bomb was dropped on Nagasaki and the war was over. So far as the Japanese could know, our ability to execute atomic attacks, if necessary by many planes at a time, was unlimited. As Dr. Karl Compton has said, "it was not one atomic bomb, or two, which brought surrender; it was the experience of what an atomic bomb will actually do to a community, plus the dread of many more, that was effective." ...

A PERSONAL SUMMARY

In the foregoing pages I have tried to give an accurate account of my own personal observations of the circumstances which led up to the use of the atomic bomb and the reasons which underlay our use of it. To me they have always seemed compelling and clear, and I cannot see how any person vested with such responsibilities as mine could have taken any other course or given any other advice to his chiefs.

Two great nations were approaching contact in a fight to a finish which would begin on November 1, 1945. Our enemy, Japan, commanded forces of somewhat over 5,000,000 armed men. Men of these armies had already inflicted upon us, in our breakthrough of the outer perimeter of their defenses, over 300,000 battle casualties. Enemy armies still unbeaten had the strength to cost us a million more. As long as the Japanese government refused to surrender, we should be forced to take and hold the ground, and smash the Japanese ground armies, by close-in fighting of the same desperate and costly kind that we had faced in the Pacific islands for nearly four years. . . .

My chief purpose was to end the war in victory with the least possible cost in the lives of the men in the armies which I had helped to raise. In the light of the alternatives which, on a fair estimate, were open to us I believe that no man, in our position and subject to our responsibilities, holding in his hands a weapon of such possibilities for accomplishing this purpose and saving those lives, could have failed to use it and afterwards looked his countrymen in the face.

As I read over what I have written, I am aware that much of it, in this year of peace, may have a harsh and unfeeling sound. It would perhaps be possible to say the same things and say them more gently. But I do not think it would be wise. As I look back over the five years of my service as Secretary of War, I see too many stern and heartrending decisions to be willing to pretend that war is anything else than what it is. The face of war is the face of death; death is an inevitable part of every order that a wartime leader gives. The decision to use the atomic bomb was a decision that brought death to over a hundred thousand Japanese. No explanation can change that fact and I do not wish to gloss it over. But this deliberate, premeditated destruction was our least abhorrent choice. The destruction of Hiroshima and Nagasaki put an end to the Japanese war. It stopped the fire raids, and the strangling blockade; it ended the ghastly specter of a clash of great land armies.

In this last great action of the Second World War we were given final proof that war is death. War in the twentieth century has grown steadily more barbarous, more destructive, more debased in all its aspects.   Now, with the release of atomic energy, man's ability to destroy himself is very nearly complete.   The bombs dropped on Hiroshima and Nagasaki ended a war. They also make it wholly clear that we must never have another war. This is the lesson men and leaders everywhere must learn, and I believe that when they learn it they will find a way to lasting peace. There is no other choice.

APPENDIX 17 Numbers of Nuclear Weapons

The numbers involved in the strategic balance on which deterrence is based, 1970 and a projection for 1975.

		United States					Soviet Union			
		Number of vehicles		Number of warheads				Number of vehicles 1970	Recent incr.	Number of warheads 1970
	Vehicle	Warhead	1970	1975	1970	1975	Vehicle	Warhead			
ICBM Intercontinental ballistic missiles	Minute-Man	1 MT	490	0	490	0	SS-11	1–2 MT	800	200/yr.	800
	M-M 2	1–2 MT	500	500	500	500	SS-9	25 MT or 3 × (5 MT) MRV, not MIRV*	260	50/yr.	260
	M-M 3	3 × (1/2 MT) MIRV	10	500	30	1500					
	Titan	5 MT	54	54	54	54	SS-7, 8, 13	1–5 MT	260		260
Totals			1054	1054	1074	2054			1300		1300

Sources: *The Military Balance 1970–71*, Institute for Strategic Studies, London. *Aviation Week. Ordnance Newsletter.* "Report on Military Spending," M.C.P.T.L. R. L. Guyer, personal communication.

*That is, multiple reentry vehicles, or several warheads that scatter but are not independently targetable.

APPENDIX 17 (continued)

		United States					Soviet Union			
	Vehicle	Warhead	Number of vehicles 1970	Number of vehicles 1975	Number of warheads 1970	Number of warheads 1975	Vehicle	Warhead	Number of vehicles 1970	Recent incr.	Number of warheads 1970
SLBM Submarine-launched ballistic missiles	Pol- A-2 aris A-3 Poseidon	0.8 MT 3 × (0.2 MT) 10 × (50 KT)	13 × 16 28 × 16 0	0 10 × 16 31 × 16	203 1344 0	0 480 4960	Nuclear: H-Class Y-Class Diesel: G-Class Z-Class	1 MT 1 MT MRBM: 1 MT 1 MT	10 × 3 8 × 16 25 × 3 10 × 2	7/yr.	30 128 75 20
					1552	5440			263		263
LRB Long-range bombers	B-52 FB-111	(?)4 × (5 MT) 2 × (5 MT)	505 35	35+ ?	2020 70	2020 70	Bison Bear (+ 60 tankers)	3 × (5 MT) 3 × (5 MT)	90 50		270 150
			540		2090	2090			140		420
Grand total Tactical aircraft		5 MT	2000		4715	9584			1703		1980

The Arms Race

THE PARTIAL TEST-BAN TREATY

Excerpts from treaties and other materials

Treaty Banning Nuclear Weapon Tests in the Atmosphere,
    in Outer Space and Under Water.
Done at Moscow, August 5, 1963
Ratification advised by the U.S.A. September 24, 1963
Ratified by the President of the U.S.A. October 7, 1963
Ratifications of the Governments of the U.S.A., U.K., and U.S.S.R.
    deposited with the said Governments at Washington, London, and
    Moscow October 10, 1963
Proclaimed by the President of the U.S.A. October 10, 1963
Entered into force October 10, 1963

The Governments of the United States of America, the United Kingdom of Great Britain and Northern Ireland, and the Union of Soviet Socialist Republics, hereinafter referred to as the "Original Parties",

Proclaiming as their principal aim the speediest possible achievement of an agreement on general and complete disarmament under strict international control in accordance with the objectives of the United Nations which would put an end to the armaments race and eliminate the incentive to the production and testing of all kinds of weapons, including nuclear weapons,

Seeking to achieve the discontinuance of all test explosions of nuclear weapons for all time, determined to continue negotiations to this end, and desiring to put an end to the contamination of man's environment by radioactive substances,

Have agreed as follows:

Article I

1. Each of the Parties to this Treaty undertakes to prohibit, to prevent, and not to carry out any nuclear weapon test explosion, or any other nuclear explosion, at any place under its jurisdiction or control:

(a) in the atmosphere; beyond its limits, including outer space; or underwater, including territorial waters or high seas; or

(b) in any other environment if such explosion causes radioactive debris to be present outside the territorial limits of the State under whose jurisdiction or control such explosion is conducted. It is understood in this connection that the provisions of this subparagraph are without prejudice to the conclusion of a treaty resulting in the permanent banning of all nuclear test explosions, including all such explosions underground, the conclusion of which, as the Parties have stated in the Preamble to this Treaty, they seek to achieve.

2. Each of the Parties to this Treaty undertakes furthermore to refrain from causing, encouraging, or in any way participating in, the carrying out of any nuclear weapon test explosion, or any other nuclear explosion, anywhere which would take place in any of the environments described, or have the effect referred to, in paragraph 1 of this Article.

Article II

1. Any Party may propose amendments to this Treaty. The text of any proposed amendment shall be submitted to the Depositary Governments which shall circulate it to all Parties of this Treaty. Thereafter, if requested to do so by one-third or more of the Parties, the Depositary Governments shall convene a conference, to which they shall invite all Parties, to consider such amendment.

2. Any amendment to this Treaty must be approved by a majority of the votes of all the Parties to this Treaty, including the votes of all of the Original Parties. The amendment shall enter into force for all Parties upon the deposit of instruments of ratification by a majority of all the Parties, including the instruments of ratification of all of the Original Parties.

Article III

1. This Treaty shall be open to all States for signature. Any State which does not sign this Treaty before its entry into force in accordance with paragraph 3 of this Article may accede to it at any time.

2. This Treaty shall be subject to ratification by signatory States. Instruments of ratification and instruments of accession shall be deposited with the Governments of the Original Parties—the United States of America, the United Kingdom of Great Britain and Northern Ireland, and the Union of Soviet Socialist Republics—which are hereby designated the Depositary Governments.

3. This Treaty shall enter into force after its ratification by all the Original Parties and the deposit of their instruments of ratification.

4. For States whose instruments of ratification or accession are deposited subsequent to the entry into force of this Treaty, it shall enter into force on the date of the deposit of their instruments of ratification or accession.

5. The Depositary Governments shall promptly inform all signatory and acceding States of the date of each signature, the date of deposit of each instrument of ratification of and accession to this Treaty, the date of its entry into force, and the date of receipt of any requests for conferences or other notices.

6. This Treaty shall be registered by the Depositary Governments pursuant to Article 102 of the Charter of the United Nations.

Article IV
This Treaty shall be of unlimited duration.

Each Party shall in exercising its national sovereignty have the right to withdraw from the Treaty if it decides that extraordinary events, related to the subject matter of this Treaty, have jeopardized the supreme interests of its country. It shall give notice of such withdrawal to all other Parties to the Treaty three months in advance.

Article V
[The final Article V is merely procedural.]

TREATY ON THE NON-PROLIFERATION OF NUCLEAR WEAPONS
Done at Washington, London, and Moscow July 1, 1968;
Entered into force March 5, 1970.

The States concluding this Treaty, hereinafter referred to as the "Parties to the Treaty",

Considering the devastation that would be visited upon all mankind by a nuclear war and the consequent need to make every effort to avert the danger of such a war and to take measures to safeguard the security of peoples,

Believing that the proliferation of nuclear weapons would seriously enhance the danger of nuclear war,

In conformity with resolutions of the United Nations General Assembly calling for the conclusion of an agreement on the prevention of wider dissemination of nuclear weapons,

Undertaking to cooperate in facilitating the application of International Atomic Energy Agency safeguards on peaceful nuclear activities,

Declaring their intention to achieve at the earliest possible date the cessation of the nuclear arms race and to undertake effective measures in the direction of nuclear disarmament,

Have agreed as follows.

Article I
Each nuclear-weapon State Party to the Treaty undertakes not to transfer to any recipient whatsoever nuclear weapons or other nuclear explosive devices or control over such weapons or explosive devices directly, or indirectly; and not in any way to assist, encourage, or induce any non-nuclear-weapon State to manufacture or otherwise acquire nuclear weapons or other nuclear explosive devices, or control over such weapons or explosive devices.

Article II
Each non-nuclear-weapon State Party to the Treaty undertakes not to receive the transfer from any transferor whatsoever of nuclear weapons or other nuclear explosive devices or of control over such weapons or explosive devices directly, or indirectly; not to manufacture or otherwise acquire nuclear weapons or other nuclear explosive devices; and not to seek or receive any assistance in the manufacture of nuclear weapons or other nuclear explosive devices.

Article III
1. Each non-nuclear-weapon State Party to the Treaty undertakes to accept safeguards, as set forth in an agreement to be negotiated and concluded with the International Atomic Energy Agency and the Agency's safeguards system, for the exclusive purpose of verification of the fulfillment of its obligations assumed under this Treaty with a

view to preventing diversion of nuclear energy from peaceful uses to nuclear weapons or other nuclear explosive devices. Procedures for the safeguards required by this article shall be followed with respect to source or special fissionable material whether it is being produced, processed or used in any principal nuclear facility or is outside any such facility. The safeguards required by this article shall be applied on all source or special fissionable material in all peaceful nuclear activities within the territory of such State, under its jurisdiction, or carried out under its control anywhere.

2. Each State Party to the Treaty undertakes not to provide: (a) source or special fissionable material, or (b) equipment or material especially designed or prepared for the processing, use or production of special fissionable material, to any non-nuclear-weapon State for peaceful purposes, unless the source of special fissionable material shall be subject to the safeguards required by this article.

3. The safeguards required by this article shall be implemented in a manner designed to comply with article IV of this treaty, and to avoid hampering the economic or technological development of the Parties or international cooperation in the field of peaceful nuclear activities, including the international exchange of nuclear material and equipment for the processing, use or production of nuclear material for peaceful purposes in accordance with the provisions of this article and the principle of safeguarding set forth in the Preamble of the Treaty.

4. Non-nuclear-weapon States Party to the Treaty shall conclude agreements with the International Atomic Energy Agency to meet the requirements of this article either individually or together with other States in accordance with the Statute of the International Atomic Energy Agency. Negotiation of such agreements shall commence within 180 days from the original entry into force of this Treaty. For States depositing their instruments of ratification or accession after the 180-day period, negotiation of such agreements shall commence not later than the date of such deposit. Such agreements shall enter into force not later than eighteen months after the date of initiation of negotiations.

Article IV

1. Nothing in this Treaty shall be interpreted as affecting the inalienable right of all the Parties to the Treaty to develop research, production and

use of nuclear energy for peaceful purposes without discrimination and in conformity with articles I and II of this Treaty.

2. All the Parties to the Treaty undertake to facilitate, and have the right to participate in, the fullest possible exchange of equipment, materials and scientific and technological information for the peaceful uses of nuclear energy. Parties to the Treaty in a position to do so shall also cooperate in contributing alone or together with other States or international organizations to the further development of the applications of nuclear energy for peaceful purposes, especially in the territories of non-nuclear-weapon States Party to the Treaty, with due consideration for the needs of the developing areas of the world.

Article V
Each Party to the Treaty undertakes to take appropriate measures to ensure that, in accordance with this Treaty, under appropriate international observation and through appropriate international procedures, potential benefits from any peaceful applications of nuclear explosions will be made available to non-nuclear-weapon States Party to the Treaty on a non-discriminatory basis and that the charge to such Parties for the explosive devices used will be as low as possible and exclude any charge for research and development. Non-nuclear-weapon States Party to the Treaty shall be able to obtain such benefits, pursuant to a special international agreement or agreements, through an appropriate international body with adequate representation of non-nuclear-weapon States. Negotiations on this subject shall commence as soon as possible after the Treaty enters into force. Non-nuclear-weapon States Party to the Treaty so desiring may also obtain such benefits pursuant to bilateral agreements.

Article VI
Each of the Parties to the Treaty undertakes to pursue negotiations in good faith on effective measures relating to cessation of the nuclear arms race at an early date and to nuclear disarmament, and on a treaty on general and complete disarmament under strict and effective international control.

Article VII
Nothing in this Treaty affects the right of any group of States to conclude regional treaties in order to assure the total absence of nuclear weapons in their respective territories.

Article VIII

1. Any Party to the Treaty may propose amendments to this Treaty. The text of any proposed amendment shall be submitted to the Depositary Governments which shall circulate it to all Parties to the Treaty. Thereupon, if requested to do so by one-third or more of the Parties to the Treaty, the Depositary Governments shall convene a conference, to which they shall invite all the Parties to the Treaty, to consider such an amendment.

    2. Any amendment to this Treaty must be approved by a majority of the votes of all the Parties to the Treaty, including the votes of all nuclear-weapon States Party to the Treaty and all other parties which, on the date the amendment is circulated, are members of the Board of Governors of the International Atomic Energy Agency. The amendment shall enter into force for each Party that deposits its instrument of ratification of the amendment upon the deposit of such instruments of ratification by a majority of all the Parties, including the instruments of ratification of all nuclear-weapon States Party to the Treaty and all other Parties which, on the date the amendment is circulated, are members of the Board of Governors of the International Atomic Energy Agency. Thereafter, it shall enter into force for any other Party upon the deposit of its instrument of ratification of the amendment.

    3. Five years after the entry into force of this Treaty, a conference of Parties to the Treaty shall be held in Geneva, Switzerland, in order to review the operation of this Treaty with a view to assuring that the purposes of the Preamble and the provisions of the Treaty are being realized. At intervals of five years thereafter, a majority of the Parties to the Treaty may obtain, by submitting a proposal to this effect to the Depositary Governments, the convening of further conferences with the same objective of reviewing the operation of the Treaty.

Article IX

1. This Treaty shall be open to all States for signature. Any State which does not sign the Treaty before its entry into force in accordance with paragraph 3 of this article may accede to it at any time.

    2. This Treaty shall be subject to ratification by signatory States. Instruments of ratification and instruments of accession shall be deposited with the Governments of the United States of America, the United Kingdom of Great Britain and Northern Ireland and the Union

of Soviet Socialist Republics, which are hereby designated the Depositary Governments.

3. This Treaty shall enter into force after its ratification by the States, the Governments of which are designated Depositaries of the Treaty, and forty other States signatory to this Treaty and the deposit of their instruments of ratification. For the purposes of this Treaty, a nuclear-weapon State is one which has manufactured and exploded a nuclear weapon or other nuclear explosive device prior to January 1, 1967.

4. For States whose instruments of ratification or accession are deposited subsequent to the entry into force of this Treaty, it shall enter into force on the date of the deposit of their instruments of ratification or accession.

5. The Depositary Governments shall promptly inform all signatory and acceding States of the date of each signature, the date of deposit of each instrument of ratification or of accession, the date of the entry into force of this Treaty, and the date of receipt of any requests for convening a conference or other notices.

6. This Treaty shall be registered by the Depositary Governments pursuant to article 102 of the Charter of the United Nations.

Article X

1. Each Party shall in exercising its national sovereignty have the right to withdraw from the Treaty if it decides that extraordinary events, related to the subject matter of this Treaty, have jeopardized the supreme interests of its country. It shall give notice of such withdrawal to all other Parties to the Treaty and to the United Nations Security Council three months in advance. Such notice shall include a statement of the extraordinary events it regards as having jeopardized its supreme interests.

2. Twenty-five years after the entry into force of the Treaty, a conference shall be convened to decide whether the Treaty shall continue in force indefinitely, or shall be extended for an additional fixed period or periods. This decision shall be taken by a majority of the Parties to the Treaty.

REMARKS BY SECRETARY OF DEFENSE ROBERT S. McNAMARA

Addressed to United Press International editors and publishers
San Francisco, California, September 18, 1967

I want to discuss with you this afternoon the gravest problem that an American Secretary of Defense must face: the planning, preparation and policy governing the possibility of thermonuclear war.

One must begin with precise definitions.

The cornerstone of our strategic policy continues to be to deter deliberate nuclear attack upon the United States, or its allies, by maintaining a highly reliable ability to inflict an unacceptable degree of damage upon any single aggressor, or combination of aggressors, at any time during the course of a strategic nuclear exchange—even after our absorbing a surprise first strike.

This can be defined as our "assured destruction capability."

Now it is imperative to understand that assured destruction is the very essence of the whole deterrence concept.

The point is that a potential aggressor must himself believe that our assured destruction capability is in fact actual, and that our will to use it in retaliation to an attack is in fact unwavering.

The conclusion, then, is clear: if the United States is to deter a nuclear attack on itself or on our allies, it must possess an actual, and a credible assured destruction capability.

When calculating the force we require, we must be "conservative" in all our estimates of both a potential aggressor's capabilities, and his intentions.   Security depends upon taking a "worst plausible case" —and having the ability to cope with that eventuality.

In that eventuality, we must be able to absorb the total weight of nuclear attack on our country—on our strike-back forces; on our command and control apparatus; on our industrial capacity; on our cities; and on our population—and still be fully capable of destroying the aggressor to the point that his society is simply no longer viable in any meaningful twentieth-century sense.

That is what deterrence to nuclear aggression means. It means the certainty of suicide to the aggressor—not merely to his military forces, but to his society as a whole.

Now let us consider another term: "first-strike capability." This, in itself, is an ambiguous term, since it could mean simply the ability of one nation to attack another nation with nuclear forces first. But as it is normally used, it connotes much more: the substantial elimination of the attacked nation's retaliatory second-strike forces.

This is the sense in which "first-strike capability" should be understood.

Now, clearly, such a first-strike capability is an important strategic concept. The United States cannot—and will not—ever permit itself to get into the position in which another nation, or combination of nations, would possess such a first-strike capability, which could be effectively used against it.

To get into such a position vis-à-vis any other nation or nations would not only constitute an intolerable threat to our security, but it would obviously remove our ability to deter nuclear aggression—both against ourselves and against our allies.

Now, we are not in that position today—and there is no foreseeable danger of our ever getting into that position.

Our strategic offensive forces are immense: 1000 Minutemen missile launchers, carefully protected below ground, 41 Polaris submarines, carrying 656 missile launchers—with the majority of these hidden beneath the seas at all times; and about 600 long-range bombers, approximately forty percent of which are kept always in a high state of alert.

Our alert forces alone carry more than 2200 weapons, averaging more than one megaton each. A mere 400 one-megaton weapons, if delivered on the Soviet Union, would be sufficient to destroy over one-third of her population, and one-half of her industry. And all of these flexible and highly reliable forces are equipped with devices that insure their penetration of Soviet defenses.

Now what about the Soviet Union?

Does it today possess a powerful nuclear arsenal?

The answer is that it does.

Does it possess a first-strike capability against the United States?

The answer is that it does not.

Can the Soviet Union, in the foreseeable future, acquire such a first-strike capability against the United States?

The answer is that it cannot.

It cannot because we are determined to remain fully alert, and we will never permit our own assured destruction capability to be at a point where a Soviet first-strike capability is even remotely feasible.

Is the Soviet Union seriously attempting to acquire a first-strike capability against the United States?

Although this is a question we cannot answer with absolute certainty, we believe the answer is no. In any event, the question itself is —in a sense—irrelevant. It is irrelevant since the United States will so continue to maintain—and where necessary strengthen—our retaliatory forces, that whatever the Soviet Union's intentions or actions, we will continue to have an assured destruction capability vis-à-vis their society in which we are completely confident.

But there is another question that is most relevant.

And that is, do we—the United States—possess a first-strike capability against the Soviet Union?

The answer is that we do not.

And we do not, not because we have neglected our nuclear strength. On the contrary, we have increased it to the point that we possess a clear superiority over the Soviet Union.

We do not possess first-strike capability against the Soviet Union for precisely the same reason that they do not possess it against us.

And that is that we have both built up our "second-strike capability"* to the point that a first-strike capability on either side has become unattainable.

For the most meaningful and realistic measurement of nuclear capability is neither gross megatonnage, nor the number of available missile launchers; but rather the number of separate warheads that are capable of being delivered with accuracy on individual high-priority targets with sufficient power to destroy them.

Gross megatonnage in itself is an inadequate indicator of assured destruction capability, since it is unrelated to survivability, accuracy, or penetrability, and poorly related to effective elimination of multiple high-priority targets.   There is manifestly no advantage in over-destroying one target, at the expense of leaving undamaged other targets of equal importance.

Further, the number of missile launchers available is also an inadequate indicator of assured destruction capability, since the

---

* A second-strike capability is the capability to absorb a surprise nuclear attack and survive with sufficient power to inflict unacceptable damage on the aggressor.

fact is that many of our launchers will carry multiple warheads.

But by using the realistic measurement of the number of warheads available, capable of being reliably delivered with accuracy and effectiveness on the appropriate targets in the United States or Soviet Union, I can tell you that the United States currently possesses a superiority over the Soviet Union of at least three or four to one.

What is important to understand is that our nuclear strategic forces play a vital and absolutely necessary role in our security and that of our allies, but it is an intrinsically limited role.

Thus, we and our allies must maintain substantial conventional forces, fully capable of dealing with a wide spectrum of lesser forms of political and military aggression—a level of aggression against which the use of strategic nuclear forces would not be to our advantage, and thus a level of aggression which these strategic nuclear forces by themselves cannot effectively deter. One cannot fashion a credible deterrent out of an incredible action. Therefore security for the United States and its allies can only arise from the possession of a whole range of graduated deterrents, each of them fully credible in its own context.

Now I have pointed out that in strategic nuclear matters, the Soviet Union and the United States mutually influence one another's plans.

In recent years the Soviets have substantially increased their offensive forces. We have, of course, been watching and evaluating this very carefully.

Clearly, the Soviet build-up is in part a reaction to our own build-up since the beginning of this decade.

Soviet strategic planners undoubtedly reasoned that if our build-up were to continue at its accelerated pace, we might conceivably reach, in time, a credible first-strike capability against the Soviet Union.

That was not in fact our intention. Our intention was to assure that they—with their theoretical capacity to reach such a first-strike capability—would not in fact outdistance us.

But they could not read our intentions with any greater accuracy than we could read theirs. And thus the result has been that we have both built up our forces to a point that far exceeds credible second-strike capability against the forces we each started with.

In doing so, neither of us has reached a first-strike capability. And the realities of the situation being what they are—whatever we believe their intentions to be, and whatever they believe our intentions to be

—each of us can deny the other a first-strike capability in the foreseeable future.

We do not want a nuclear arms race with the Soviet Union—primarily because the action-reaction phenomenon makes it foolish and futile. But if the only way to prevent the Soviet Union from obtaining first-strike capability over us is to engage in such a race, the United States possesses in ample abundance the resources, the technology, and the will to run faster in that race for whatever distance is required.

But what we would much prefer to do is to come to a realistic and reasonably riskless agreement with the Soviet Union, which would effectively prevent such an arms race. We both have strategic nuclear arsenals greatly in excess of a credible assured destruction capability. These arsenals have reached that point of excess in each case for precisely the same reason: we each have reacted to the other's build-up with very conservative calculations. We have, that is, each built a greater arsenal than either of us needed for a second-strike capability, simply because we each wanted to be able to cope with the "worst plausible case."

But since we now each possess a deterrent in excess of our individual needs, both of our nations would benefit from a properly safeguarded agreement first to limit, and later to reduce, both our offensive and defensive strategic nuclear forces.

We may, or we may not, be able to achieve such an agreement. We hope we can. And we believe such an agreement is fully feasible, since it is clearly in both our nations' interests.

The Soviets are now deploying an anti-ballistic missile system. If we react to this deployment intelligently, we have no reason for alarm.

The system does not impose any threat to our ability to penetrate and inflict massive and unacceptable damage on the Soviet Union. In other words, it does not presently affect in any significant manner our assured destruction capability.

It does not impose such a threat because we have already taken the steps necessary to assure that our land-based Minuteman missiles, our nuclear submarine-launched new Poseidon missiles, and our strategic bomber forces have the requisite penetration aids—and in the sum, constitute a force of such magnitude, that they guarantee us a force strong enough to survive a Soviet attack and penetrate the Soviet ABM deployment.

Now let me come to the issue that has received so much attention recently: the question of whether or not we should deploy an ABM system against the Soviet nuclear threat.

To begin with, this is not in any sense a new issue. We have had both the technical possibility and the strategic desirability of an American ABM deployment under constant review since the late 1950s.

While we have substantially improved our technology in the field, it is important to understand that none of the systems at the present or foreseeable state of the art would provide an impenetrable shield over the United States. Were such a shield possible, we would certainly want it—and we would certainly build it.

In point of fact, we have already initiated offensive weapons programs costing several billions in order to offset the small present Soviet ABM deployment, and the possibly more extensive future Soviet ABM deployments.

That is money well spent; and it is necessary.

But we should bear in mind that it is money spent because of the action-reaction phenomenon.

If we in turn opt for heavy ABM deployment—at whatever price —we can be certain that the Soviets will react to offset the advantage we would hope to gain.

It is precisely because of this certainty of a corresponding Soviet reaction that the four prominent scientists—men who have served with distinction as the Science Advisors to Presidents Eisenhower, Kennedy, and Johnson—and the three outstanding men who have served as Directors of Research and Engineering to three Secretaries of Defense have unanimously recommended against the deployment of an ABM system designed to protect our population against a Soviet attack.

These men are Doctors Killian, Kistiakowsky, Wiesner, Hornig, York, Brown, and Foster.

There is evidence that the Chinese are devoting very substantial resources to the development of both nuclear warheads and missile delivery systems. As I stated last January, indications are that they will have medium-range ballistic missiles within a year or so, an initial intercontinental ballistic missile capability in the early 1970s, and a modest force in the mid-70s.

We possess now, and will continue to possess for as far ahead as we can foresee, an overwhelming first-strike capability with respect to

China. And despite the shrill and raucous propaganda directed at her own people that "the atomic bomb is a paper tiger," there is ample evidence that China well appreciates the destructive power of nuclear weapons.

China has been cautious to avoid any action that might end in a nuclear clash with the United States—however wild her words—and understandably so. We have the power not only to destroy completely her entire nuclear offensive forces, but to devastate her society as well.

Is there any possibility, then, that by the mid-1970s China might become so incautious as to attempt a nuclear attack on the United States or our allies?

It would be insane and suicidal for her to do so, but one can conceive conditions under which China might miscalculate. We wish to reduce such possibilities to a minimum.

And since, as I have noted, our strategic planning must always be conservative, and take into consideration even the possible irrational behavior of potential adversaries, there are marginal grounds for concluding that a light deployment of United States ABMs against this possibility is prudent.

Moreover, such an ABM deployment designed against a possible Chinese attack would have a number of other advantages. It would provide an additional indication to Asians that we intend to deter China from nuclear blackmail, and thus would contribute toward our goal of discouraging nuclear weapon proliferation among the present non-nuclear countries.

Further, the Chinese-oriented ABM deployment would enable us to add—as a concurrent benefit—a further defense of our Minuteman sites against Soviet attack, which means that at modest cost we would in fact be adding even greater effectiveness to our offensive missile force and avoiding a much more costly expansion of that force.

Finally, such a reasonably reliable ABM system would add protection of our population against the improbable but possible accidental launch of an intercontinental missile by any one of the nuclear powers.

After a detailed review of all these considerations, we have decided to go forward with this Chinese-oriented ABM deployment, and we will begin actual production of such a system at the end of this year.

In reaching this decision, I want to emphasize that it contains two possible dangers—and we should guard carefully against each.

The first danger is that we may psychologically lapse into the over-simplification about the adequacy of nuclear power. The simple truth is that nuclear weapons can serve to deter only a narrow range of threats. This ABM deployment will strengthen our defensive posture—and will enhance the effectiveness of our land-based ICBM offensive forces. But the independent nations of Asia must realize that these benefits are no substitute for their maintaining, and where necessary strengthening, their own conventional forces in order to deal with the more likely threats to the security of the region.

The second danger is also psychological. There is a kind of mad momentum intrinsic to the development of all new nuclear weaponry. If a weapon system works—and works well—there is strong pressure from many directions to procure and deploy the weapon out of all proportion to the prudent level required.

The danger in deploying this relatively light and reliable Chinese-oriented ABM system is going to be that pressures will develop to expand it into a heavy Soviet-oriented ABM system.

We must resist that temptation firmly—not because we can for a moment afford to relax our vigilance against a possible Soviet first-strike—but precisely because our greatest deterrent against such a strike is not a massive, costly, but highly penetrable ABM shield, but rather a fully credible offensive assured destruction capability.

The so-called heavy ABM shield—at the present state of technology —would in effect be no adequate shield at all against a Soviet attack, but rather a strong inducement for the Soviets to vastly increase their own offensive forces. That, as I have pointed out, would make it necessary for us to respond in turn—and so the arms race would rush hopelessly on to no sensible purpose on either side.

Let me emphasize—and I cannot do so too strongly—that our decision to go ahead with a limited ABM deployment in no way indicates that we feel an agreement with the Soviet Union on the limitation of strategic nuclear offensive and defensive forces is any the less urgent or desirable.

The road leading from the stone axe to the ICBM—though it may have been more than a million years in the building—seems to have run in a single direction.

If one is inclined to be cynical, one might conclude that man's history seems to be characterized not so much by consistent periods of peace, occasionally punctuated by warfare; but rather by persistent outbreaks of warfare, wearily put aside from time to time by periods of exhaustion and recovery—that parade under the name of peace.

I do not view man's history with that degree of cynicism, but I do believe that man's wisdom is avoiding war is often surpassed by his folly in promoting it.

However foolish unlimited war may have been in the past, it is now no longer merely foolish, but suicidal as well.

It is said that nothing can prevent a man from suicide, if he is sufficiently determined to commit it.

The question is what is our determination in an era when unlimited war will mean the death of hundreds of millions—and the possible genetic impairment of a million generations to follow?

Man is clearly a compound of folly and wisdom—and history is clearly a consequence of the admixture of those two contradictory traits. History has placed our particular lives in an era when the consequences of human folly are waxing more and more catastrophic in the matters of war and peace.

In the end, the root of man's security does not lie in his weaponry.

In the end, the root of man's security lies in his mind.

References and Bibliography

Preface

1. Phillip Sporn, *Energy*. Elmsford, N.Y., Pergamon Press, 1963.
2. M. King Hubbert, "Energy Resources of the Earth," *Sci. Amer.*, Sept. 1971, p. 61.
3. Chauncey Starr, "Energy and Power," *Sci. Amer.*, Sept. 1971, p. 37.

Chapter 1

1. Ira M. Freeman, *Physics Made Simple*. New York, Macmillan, 1964.
2. C. E. Bennett, *Physics without Mathematics*. New York, Barnes and Noble, 1949.

Chapter 2

1. R. E. Lapp, "The Nuclear Power Controversy—Power and Hot Water," *New Republic*, Feb. 6, 1971, p. 20.

Chapter 3

1. V. H. Booth, *The Structure of Atoms*. New York, Macmillan, 1964.
2. G. Gamow, *The Atom and Its Nucleus*. Englewood Cliffs, N.J., Prentice Hall, 1961.
3. R. Adler, *Inside the Nucleus*. New York, John Day, 1963.
4. Gregory Choppin, *Nuclei and Radioactivity*. New York, W. A. Benjamin, 1964.
5. H. G. Graetzer and D. L. Anderson, *The Discovery of Nuclear Fission*. New York, Van Nostrand Reinhold, 1971.
6. D. R. Inglis, "Nuclear Models," *Physics Today*, June 1970.
7. "Nucleus, Atomic," *Encyclopaedia Britannica*, 1973 ed.

Chapter 4

1. R. L. Lyerly, "Nuclear Power Plants," *AEC Understanding the Atom* series.

2. J. F. Edgerton, "Nuclear Reactors," *AEC Understanding the Atom* series.

3. G. T. Seaborg and W. R. Corliss, *Man and Atom*, Dutton, New York (1971). G. T. Seaborg and J. L. Bloom, "Fast Breeder Reactors," *Sci. Amer.*, Nov. 1970. See Appendix 15.

4. A. P. Bray, "Basic Information about Reactors," in *Nuclear Reactors*, H. Foreman, ed. Minneapolis, Univ. Minnesota Press, 1970.

5. D. H. Hughes, *On Nuclear Energy*. Cambridge, Harvard Univ. Press, 1957.

6. S. Glasstone, *Sourcebook on Atomic Energy*. Princeton, Van Nostrand, 1967.

7. S. Novick, *The Careless Atom*. Boston, Houghton Mifflin, 1969.

8. S. Novick, "Continuing the Fermi Story," *Scientist and Citizen* **9**, 9, 228 (1968). Also S. Novick, "A Mile from Times Square," *Environment* **11**, 1, 10, 1969. [The emergency core-cooling problem.]

9. T. J. Thompson, "Accidents and Destructive Tests," in *The Technology of Nuclear Reactor Safety*, T. J. Thompson and J. G. Beckerly, eds., Vol. I, p. 608. Cambridge, M.I.T. Press, 1964.

10. "Costs and Benefits of Nuclear Electric Power Plants," Committee for Environmental Information (Minneapolis, Box 14026, Univ. Station), 1969.

11. R. E. Lapp, "The Nuclear Plant Controversy—Safety," *New Republic*, Jan. 23, 1971, p. 18.

12. W. H. Jordan, "Nuclear Energy: Benefits versus Risks," *Physics Today*, May 1970, p. 32.

13. D. R. Inglis and G. R. Ringo, "Underground Construction of Nuclear Reactors," Argonne National Laboratory Report ANL-5652 (1957). Also F. C. Rogers, "Underground Nuclear Power Plants," *Bull. Atomic Scientists*, Oct. 1971, p. 38.

14. I. A. Forbes, D. F. Ford, H. W. Kendall, and J. J. Mackenzie, "Cooling Water," *Environment* **14**, 41, Jan. 1972; also "Nuclear Reactor Safety, an Evaluation of New Evidence," and "A Critique of the New AEC Design Criteria for Reactor Safety Systems." Union of Concerned Scientists, P.O. Box 299, Cambridge, Mass., 02139, or Congressional Record 117, 154, S. 16364.

15. F. R. Farmer, "Safety of Nuclear Power Plants: A British View," *Bull. Atomic Scientists*, Nov. 1971, p. 47.

16. V. S. Emelyanov, "Nuclear Energy in the Soviet Union," *Bull. Atomic Scientists*, Nov. 1971, p. 38.

Chapter 5

1. Linus Pauling, *No More War*. New York, Dodd, Mead, 1958.

2. Herman V. Muller, "Genetic Damage Produced by Radiation," in *The Atomic Age*, M. Grodzins and E. Rabinowitch, eds., p. 297. New York, Basic Books, 1963 (from *Bull. Atomic Scientists*, June 1955).

3. Herman J. Muller, "Radiation and Human Mutation," *Sci. Amer.* **193**, 20, 58, November 1955. Offprint available, San Francisco, W. H. Freeman.

4. *Fallout*, John M. Fowler, ed. New York, Basic Books, 1960. Includes W. M. Gould, "Biological Effects of Radiation," J. F. Crow, "Radiation and Future Generations," and Jack Schubert, "Protection and Treatment."

5. J. Schubert and R. A. Lapp, *Radiation and How It Affects You*. New York, Viking Press, 1966.

6. I. Asimov and T. Dobzhansky, "Genetic Effects of Radiation," *AEC Understanding the Atom* series, 1966. In the same series, N. A. Frigerio, "Your Body and Radiation." Also "Radiation in Medicine" and "Radioisotopes in Industry."

7. *Principles of Radiation Protection*, K. Z. Morgan and J. E. Turner, eds. New York, John Wiley, 1967.

8. J. J. W. Baker and G. E. Allen, *Matter, Energy, and Life*. Reading, Mass., Addison-Wesley, 1970.

9. A. R. Tamplin and J. W. Gofman, *"Population Control"! through Nuclear Pollution*. Chicago, Nelson-Hall, 1970. Also J. W. Gofman and A. R. Tamplin, *The Case Against Nuclear Power Plants*. Emmanaus, Pa., Rodale Press, 1971.

10. E. J. Sternglass, *Low-Level Radiation*, New York, Ballantine, 1972; also A. R. Hofmann and D. R. Inglis, "Radiation and Infants," *Bull. Atomic Scientists*, Sept. 1972. E. J. Sternglass, "Infant Mortality and Nuclear Tests," *Bull. Atomic Scientists* **25**, 4, 18 (April 1969); also A. Nilsson and G. Walinder, "Strontium-90 Dosages and Infant Mortality," and a reply, *Bull. Atomic Scientists* **26**, 5, 40 (May 1970).

11. Alice Steward and G. W. Kneale, "Radiation Dose Effects in Relation to Obstetric X-Rays and Childhood Cancer," *Lancet* **1**, no. 7658, p. 1185 (June 1970). Also M. P. Finkel, B. O. Biskis and P. J. Jenkins, "Toxicity of Radium-226 in Mice," *International Atomic Energy Symposium Report* IAEA-SM-118/11, p. 369 and esp. p. 387. [Evidence for linear relation to low doses.]

12. R. E. Lapp, "The Nuclear Plant Controversy—Radiation Risks," *New Republic*, Feb. 27, 1971, p. 17. Also D. Farney, "Atom Age Trash," *Wall Street J.*, Jan. 25, 1971.

13. P. M. Boffey, "Radiation Standards: Are the Right People Making the Decisions?" *Science* **171**, 780 (Feb. 26, 1971).

14. J. Snow, "Radioactive Wastes from Reactors," *Scientist and Citizen* **9**, 5, 88 (1968).

15. E. Albone and J. McCaull, "Freighted with Hazard," *Environment* **12**, 12, 18 (1970).

16. D. R. Inglis, "Nuclear Pollution and the Arms Race," *Progressive*, April 1970; also in *The Crisis of Survival.* Glenview, Ill., Scott Foresman, 1970.

17. R. S. Lewis, "Radioactive Salt Mine," *Bull. Atomic Scientists* **27**, 6, 27 (June 1971).

18. E. A. Martell et al., "Fire Damage" [Pu processing plant near Denver], *Environment* **12**, 4, 14 (1970).

19. J. A. Lieberman, "Ionizing-radiation Standards for Population Exposure," *Physics Today*, Nov. 1971, p. 32. [Suggests the probable approach of the new Environmental Protection Agency.]

20. Several articles in *Bull. Atomic Scientists*, Sept. 1970: A. R. Tamplin and J. W. Gofman, "Radiation Effects Controversy," p. 2; Linus Pauling, "Genetic and Somatic Effects of High-Energy Radiation," p. 3; T. J. Thompson and W. R. Bibb, "The AEC Position," p. 9;
    From *Bull. Atomic Scientists*, Sept. 1971, the following: P. J. Lindop and J. Rotblat, "Radiation Pollution and the Environment," p. 17; A. R. Tamplin, "Issues in the Radiation Controversy," p. 25; J. W. Gofman, "Nuclear Power and Ecocide," p. 28; Joshua Ledenberg, "Squaring an Infinite Circle: Radiobiology and the Value of Life," p. 43.

21. Roger Rapoport, *The Great American Bomb Machine.* New York, E. P. Dutton, 1971. Also R. H. Romer, "Resource Letter on Energy," *Amer. J. Physics* **40**, 805, June 1972.

22. *Biological Implications of the Nuclear Age*, AEC symposium series No. 16 (CONF-690303), 1969.

Chapter 6

1. B. W. Sharpe, "Nuclear Safeguards: The IAEA Program," *Physics Today*, Nov. 1969, p. 33.

2. "Institute for Nuclear Materials Management," *Proceedings of the Eleventh Annual Meeting*, May 25-27, 1970. Obtainable from INMM, Box 273, Argonne, Ill. 60439. Excerpts from remarks at this meeting by Congressman Craig Hosmer appear in Appendix 15.

3. D. I. Bolef, "Nuclear Weapons (A U.N. View)," *Scientist and Citizen* **10**, 2, 38 (1968).

Chapter 7

1. R. F. Post, "Controlled Fusion Research," *Ann. Rev. Nuclear Sci.* **20**, 509 (1970). [See Appendix 15.] R. F. Post, "The Prospects of Fusion Power," *Bull. Atomic Scientists*, Oct. 1971, p. 42; W. C. Gough and B. J. Eastlund, "The Prospects of Fusion Power," *Sci. Amer.*, Feb. 1971, p. 50.

2. D. F. Anthrop, "Environmental Side Effects of Energy Production," *Bull. Atomic Scientists*, Oct. 1970, p. 39.

3. A. M. Weinberg, "Nuclear Energy and the Environment," *Bull. Atomic Scientists*, June 1970, p. 69. See also the *Bulletin* for March 1971, and June 1971, p. 2. Also "State of the Laboratory 1971," *ORNL Review*, Winter 1972, and "Social Institutions and Nuclear Energy," *Science* **177**, 27 (July 1972).

4. D. R. Inglis, "Nuclear Energy and the Malthusian Dilemma," *Bull. Atomic Scientists*, Feb. 1971, p. 15, and June 1971, p. 39.

5. R. E. Lapp, "Where Will We Get the Energy?" *New Republic*, July 11, 1970, p. 17.

6. "New Boom for an Old Industry [coal]," *U.S. News and World Report*, Mar. 29, 1971, p. 60.

7. Several articles from *Bull. Atomic Scientists*, Sept. 1971: Bernard I. Spinrad, "America's Energy Crisis, Reality or Hysteria?" p. 3; Manson Benedict, "Electric Power from Nuclear Fission," p. 8; D. P. Geesaman, "Plutonium and the Energy Decision," p. 33; Glenn T. Seaborg, "On Misunderstanding the Atom," p. 46.

8. Several articles from *Bull. Atomic Scientists*, Oct. 1971: V. L. Parsegian, "New Goals for Atomic Energy," p. 2; S. D. Freeman, "Towards a Policy of Energy Conservation," p. 8; I. L. White, "Energy Policy Making: Limitations of a Conceptual Model," p. 20; N. C. Ford and J. W. Kane, "Solar Power," p. 27; A. B. and M. P. Meinel, "Is It Time for a New Look at Solar Energy?" p. 32; W. Rex, "Geothermal Energy," p. 52.

9. Alvin M. Weinberg, "Nuclear Energy: A Prelude to H. G. Wells' Dream?" *Foreign Affairs* **49**, 407 (1971).

10. A. L. Hammond, "Breeder Reactors, Power for the Future," *Science* **174**, 807 (Nov. 1971).

Chapter 8

1. H. D. Smyth, *Atomic Energy for Military Purposes*. [The "Smyth Report"] Princeton Univ. Press, 1945.

2. S. Glasstone, "The Effects of Nuclear Weapons," U.S.G.P.O., 1964.

3. R. E. Lapp, *Kill and Overkill*. New York, Basic Books, 1962.

4. R. E. Lapp, *The Weapons Culture*. New York, W. W. Norton, 1968.

5. "Defense in the Nuclear Age," *Scientist and Citizen*, Feb.–Mar. 1966. [Entire issue on shelters and civil defense.]

6. *Fallout*, John M. Fowler, ed. New York, Basic Books, 1960.

7. S. Novick, "Do It Yourself," and particularly the reply by G. Seaborg, *Environment* **11**, 10, 22 (1969). [On clandestine bombs.]

8. R. E. Lapp, *The Voyage of the Lucky Dragon*. New York, Harper, 1958. Also N. O. Hines, *Proving Ground*. Seattle, Univ. of Washington Press, 1962.

9. Flora Lewis, "Men Who Handle Nuclear Weapons Also Using Drugs," *Boston Globe*, Sept. 6, 1971.

10. C. Hohenemser and M. Seitenberg, "A Comprehensive Nuclear Test Ban, Technological Aspects 1957–67," *Scientist and Citizen* **9**, 9, 197 (1967).

11. "Earthquakes and Explosions," *Scientist and Citizen* **9**, 9, 212 (1968).

12. "Underground Nuclear Testing," Committee for Environmental Information, *Environment* **11**, 6, 2 (1969).

13. "Announced Nuclear Detonations 1964–67," *Scientist and Citizen* **9**, 9, 210 (1967). See also *Scientist and Citizen* **10**, 2, 49 (1968).

14. "Testing the Treaty" [Observations of radioactivity in Canada and Mexico.] *Environment* **11**, 3, 10 (1969).

15. Paul Jacobs, "Precautions Are Being Taken by Those Who Know," *Atlantic Monthly*, April 1971, p. 45.

16. R. E. Lapp, *Arms Beyond Doubt* [The tyranny of weapons technology.] New York, Cowles, 1970.

17. Ralph Stavins, "Kennedy's Private War," *New York Review of Books*, July 22, 1971, p. 20. [Nuclear weapons policy.]

18. "Biological and Environmental Effects of Nuclear War," JCAE hearings, June 22–26, 1959, p. 57.

Chapter 9

1. Harrison Brown, *Must Destruction Be Our Destiny?* New York, Simon and Schuster, 1946; also *One World or None*, D. Masters and K. Way, eds. New York, Whittlesey House, 1946.

2. Philip Noel-Baker, *The Arms Race*. New York, Oceana Publications, 1958.

3. *Arms Control, Disarmament and National Security*, Donald G. Brennan, ed. New York, George Braziller, 1961.

4. *Disarmament, Its Politics and Economics*, Seymour Melman, ed. Boston, Amer. Acad. Arts and Sciences, 1962. Also T. W. Wilson, Jr., *The Great Weapons Heresy*. Boston, Houghton Mifflin, 1970.

5. *The Atomic Age*, M. Grodzins and E. Rabinowitch, eds. New York, Basic Books, 1963.

6. Leo Szilard, *The Voice of the Dolphins*. New York, Simon & Schuster, 1961.

7. Two articles from *Bull. Atomic Scientists* **17**, April 1961: L. B. Sohn, "Disarmament and Arms Control by Territories," p. 130, and D. R. Inglis, "Region by Region Disarmament," p. 19. (The latter also in *Current History* **47**, Aug. 1964, p. 88.) Also L. B. Sohn, "Zonal Disarmament and Inspection," *Bull. Atomic Scientists* **18**, Sept. 1962, p. 4; D. R. Inglis, "Region-by-Region System of Inspection and Disarmament," *J. Conflict Resolution* **9**, 187 (1965); D. R. Inglis, "Transition to Disarmament," in *America Armed*, R. A. Goldwin, ed. Chicago, Rand McNally, 1963; D. R. Inglis, "Evolving Patterns of Nuclear Disarmament Proposals," *Centennial Rev.* **6**, 445 (1962); C. C. Abt, "Progressive Zonal Inspection of Disarmament," in *Disarmament and Arms Control* Vol. 2, 1, 75. Elmsford, N.Y., Pergamon Press, 1964.

8. Jeremy Stone, *Constraining the Arms Race*. Cambridge, M.I.T. Press, 1966.

9. D. R. Inglis, "Nuclear Test Ban as a Step Toward National Security," in *The Atomic Age*, M. Grodzins and E. Rabinowitch, eds. New York, Basic Books, 1963, p. 324 (from *Bull. Atomic Scientists*, June 1956). Also D. R. Inglis, "Ban H-bomb Tests and Favor the Defense," *Bull. Atomic Scientists* **10**, p. 353 (1954).

10. D. R. Inglis, "Testing and Taming of Nuclear Weapons," New York, Public Affairs Pamphlets, 1960.

11. H. R. Myers, "Extending the Nuclear Test Ban," *Sci. Amer.* **226**, Jan. 1972, p. 3. Also H. R. Myers, "Comprehensive Test Ban—Grounds for Objections Diminish," *Science* **175**, 283 (Jan. 21, 1972); David Davies, *Seismic Methods for Monitoring Underground Explosions.* Stockholm, International Institute for Peace and Conflict Research, 1968.

12. John Cockcroft, "The Perils of Nuclear Proliferation," in *Until Peace Comes*, Nigel Calder, ed. London, Penguin Press, 1968.

13. G. W. Rathjens and G. B. Kistiakowsky, "Limitation of Strategic Arms," *Sci. Amer.* **222**, 1, 19 (Jan. 1970).

14. Herbert Scoville, "Limitation of Offensive Weapons," *Sci. Amer.* **224**, 1, 15 (Jan. 1970). Also "Missile Submarines and National Security," *Sci. Amer.* **226**, 6, June 1972, p. 15.

15. Herbert Scoville, "Danger in U.S. Nuclear Policy," *The New York Times*, Dec. 2, 1970, p. 45.

16. D. R. Inglis, "H-bombs in the Backyard," *Sat. Rev.* **51**, Dec. 11, 1968, p. 11.

17. H. F. York, *Race to Oblivion*. New York, Simon and Shuster, 1971. Also H. F. York, "Military Technology and National Security" [about ABM's], *Sci. Amer.* **221**, August 1969, p. 17.

18. Two articles from *Bull. Atomic Scientists* **26**, 1, pp. 35 and 39: A. de Volpi, "MIRV—Medusa of the Nuclear Age," and J. I. Coffey, "Soviet ABM and Arms Control."

19. G. W. Rathjens, "A Breakthrough to Arms Control?" *Bull. Atomic Scientists* **27**, 6, 4 (June 1971).

20. I. F. Stone, "Arms Talks: Theater of Delusion," and "The Kennedy Test Ban Farce," *N.Y. Rev. Books*, April 23 and May 7, 1970. Also "The Shape of Nixon's New World," N.Y. Rev. of Books, June 19, 1972.

21. A. De Volpi, "Expectations from SALT," *Bull. Atomic Scientists* **26**, 5, 6 (April 1970).

22. J. J. Stone, "When and How to Use SALT," *Foreign Affairs* **48**, p262 (1970).

23. B. T. Feld, "The Sorry History of Arms Control," *Bull. Atomic Scientists* **26**, Sept. 1970, p. 22; also B. T. Feld, "Scientists' Role in Arms Control," *Bull. Atomic Scientists* **26**, Jan. 1970, p. 2.

24. Herbert Scoville, "Verification of Nuclear Arms Limitations," *Bull. Atomic Scientists* **26**, 8, 6 (Oct. 1970).

25. J. G. Wiesner, "Arms Control: Current Proposals and Problems," *Bull. Atomic Scientists* **26**, 5, 6 (May 1970).

26. W. C. Foster, "Prospects for Arms Control," and Harold Brown, "Security Through Limitations," *Foreign Affairs* **47**, 1969, p. 413 and 422.

27. B. T. Feld, "China and the Bomb," *Bull. Atomic Scientists*, Sept. 1971, p. ii.

28. "U.S. Foreign Policy for the 1970's," Report to Congress by President Nixon, Feb. 25, 1971. [Section on Arms Control, pp. 186–198.] U.S.G.P.O., Washington, D.C. 20402.

29. P. K. H. Panofsky, "Roots of the Strategic Arms Race: Ambiguity and Ignorance," *Bull. Atomic Scientists*, June 1971, p. 15.

30. *The War Economy of the United States*, Seymour Melman, ed. New York, St. Martin's Press, 1970; also Richard F. Kaufman, *The War Profiteers*. Indianapolis, Bobbs-Merrill, 1970.

31. D. R. Inglis and C. L. Sadler, "Nonmilitary Uses of Nuclear Explosives," *Bull. Atomic Scientists* **23**, 10, 46 (Dec. 1967).

32. P. Metzger, "Project Gasbuggy and Catch-85," *The New York Times Magazine*, Feb. 22, 1970.

33. E. A. Mantell, "Plowing a Nuclear Furrow," *Environment* **11**, 3, 2 (1968).

Index

Index

About the Book

This volume provides an incisive view of both the scientific and the humanistic aspects of nuclear energy problems. Thoroughly class-tested with both science and non-science students, this material is in-tended to be used as the basis of a one-semester course, although there is ample material for a year. The overriding aim of the book is to provide the foundation for understanding both the technical and the sociopolitical sides of nuclear problems — for a reader without prior knowledge of either.

The presentation of physical principles is elementary and as brief as possible, the discussion being confined to concepts essential to an un-derstanding of nuclear reactors and weapons. The author writes for the student who has insufficient motivation toward science to want to learn it better, but who needs a basic understanding of nuclear en-ergy principles as a background for making sound sociopolitical judg-ments. The book covers with an economy but sufficiency of detail the whole subject of nuclear energy — from reactors and bombs with their potential benefits and dangers to their associated ecological ef-fects and medical and industrial spin-offs.

In addition to the basic presentation in the book's nine chapters, a wealth of supplementary material is available in eighteen appendices, divided about equally between scientific and sociopolitical subjects.

The Author

David Rittenhouse Inglis is professor of physics at the University of Massachusetts, Amherst. He received the A.B. degree from Amherst College and the D.Sc. in physics from the University of Michigan in 1931. He previously taught at Ohio State University, the University of Pittsburgh, Princeton, and Johns Hopkins University, and from 1949 to 1969 he was a senior physicist at Argonne National Labora-tory. Dr. Inglis has written or contributed to many books on atomic and nuclear physics and is a member of the American Physical Society, the Federation of American Scientists (chairman 1959-60), Phi Beta Kappa, and Sigma Xi. He is on the editorial board of the *Bulletin of the Atomic Scientists* (Science and Public Affairs) and on the board of directors of the National Committee for Nuclear Responsibility and SANE.

Cover photographs: San Onofre Nuclear Generating Station. Courtesy of Southern California Edison Company.

Foreground design by Erik Rhoades Anderson.

ADDISON-WESLEY PUBLISHING COMPANY
Reading, Massachusetts · Menlo Park, California
London · Amsterdam · Don Mills, Ontario · Sydney

ISBN 0-201-03199-X